THE CHEBYSHEV POLYNOMIALS

THEODORE J. RIVLIN

Thomas J. Watson Research Center,
IBM Corporation

A WILEY-INTERSCIENCE PUBLICATION

JOHN WILEY & SONS, New York · London · Sydney · Toronto

For my mother and father

Library of Congress Cataloging in Publication Data:

Rivlin, Theodore J.   1926–
   The Chebyshev polynomials.

   (Pure and applied mathematics)
   "A Wiley-Interscience publication."
   Bibliography: p.
   1. Chebyshev polynomials.   I.   Title.

QA404.5.R58   1974   515'.55   74–10876

ISBN 0-471-72470-X

Printed in the United States of America

10 9 8 7 6 5 4 3 2 1

PREFACE

This book has two main aims: (1) to give a survey of the most important properties of the Chebyshev polynomials and (2) to introduce some interesting areas of mathematical analysis: interpolation theory, orthogonal polynomials, approximation theory, numerical integration, numerical analysis, ergodic theory, by the example of the Chebyshev polynomials. The Chebyshev polynomial is like a fine jewel that reveals different characteristics under illumination from varying positions, and I feel that apart from its great intrinsic interest it is an ideal vehicle for giving the student a taste of these various areas.

A brief outline of the book follows. In the first chapter, after definitions and notation are presented, polynomial interpolation at the zeros and extrema of the Chebyshev polynomial is thoroughly examined. The rest of the chapter is devoted to the Chebyshev polynomials as orthogonal polynomials. The point of departure of the second chapter is the minimax property of the Chebyshev polynomial on an interval. This is seen as approximating a monomial best by lower powers or alternatively maximizing the leading coefficient of a polynomial of fixed degree and size. The former point of view leads to a brief course in the theory of best uniform approximation and the latter to an essay on maximizing linear functionals on a space of polynomials. This chapter is a much amplified version of Rivlin and Shapiro [1] ([n] refers to item n after the associated name in the references). Related material may be found in Shapiro [2]. The second part of this chapter is a prelude to the systematic and much more detailed study of similar problems in Voronovskaja [1]. One highlight of the chapter is Duffin and Schaeffer's generalization of V. A. Markov's bound on the derivative of a polynomial, which appears in a book in English for the first time here.

The Chebyshev polynomials have found extensive application in numerical analysis. One important technique in applications is the expansion of a function in a series of Chebyshev polynomials. The main theme of the third chapter is the effectiveness of the partial sums of a Chebyshev expansion of a function as approximations to the function. One of the most striking properties of the set of Chebyshev polynomials is that it is closed under functional composition. The fourth, and last, chapter focuses on this facet of the polynomials and concludes with a study of their ergodic properties.

iii

An attempt has been made to maintain a moderate pace in the exposition and to spell out many details with the hope that the book might serve as well as "leisure reading" for a broader mathematical community. More than two hundred exercises of varying degrees of difficulty have been provided. Some substantial results are broken up into chains of exercises and hints have been given for the more difficult ones.

I make no encyclopedic claims for the book's coverage of facts about Chebyshev polynomials. Among the omissions of which I am aware I particularly regret that ignorance prevented me from discussing two topics, the number theoretic aspects of the Chebyshev polynomials and applications of Chebyshev polynomials in kinematics.

It is my pleasant duty to thank many friends for helpful discussions of material in the book. In particular, I want to thank my colleagues Charles Micchelli and Roy Adler. Dr. Micchelli read a preliminary version of the manuscript and made many helpful recommendations, Dr. Adler suggested several improvements in the last chapter.

THEODORE J. RIVLIN

Chappaqua, New York
March 1974

CONTENTS

1

DEFINITIONS AND
SOME ELEMENTARY
PROPERTIES

1.1. Definition of the Chebyshev Polynomials

This book is about polynomials. Let us recall that a polynomial is a function $p(x)$ which can be written in the form

$$p(x) = a_0 + a_1 x + \cdots + a_n x^n. \qquad (1.1)$$

We shall suppose that, unless otherwise stated, $a_0, \ldots, a_n$ are real numbers and x is a real variable. If $a_n \neq 0$, then we say that p is a polynomial of degree n.† We shall often be interested in the set of polynomials whose degree does not exceed n. This set is denoted by $\mathscr{P}_n$; i.e., if $p(x) = a_0 + a_1 x + \cdots + a_k x^k$ and $k \leqslant n$, then

$$p \in \mathscr{P}_n.$$

Polynomials have many agreeable properties. They can be differentiated as often as desired for any value of x, hence can be integrated over any interval. Moreover, they are "simple" in the sense that $p(x)$ is completely specified by the $n + 1$ numbers $a_0, \ldots, a_n$. It is this finiteness that makes polynomials particularly suitable as approximations to more complicated functions.

Consider the function

$$T_n(x) = \cos n\theta, \qquad (1.2)$$

where n is a nonnegative integer, $x = \cos \theta$, and $0 \leqslant \theta \leqslant \pi$. As θ increases from 0 to π, x decreases from 1 to -1. The function $T_n(x)$ is defined by (1.2) on the interval $-1 \leqslant x \leqslant 1$, which we also denote by I; i.e., given $x \in I$, we find the

---

† Thus nonzero constants are polynomials of degree zero. Zero is in the anomalous position of being a polynomial without a degree. We remedy this situation by assigning the degree -1 to the polynomial $p = 0$.

1

unique value of θ = arc cos x which satisfies $0 \leqslant \theta \leqslant \pi$ and $T_n(x)$ has the value cos $n\theta$. Thus $T_n(x)$ is a single-valued function defined on I, which may be written

$$T_n(x) = \cos n \,(\text{arc cos } x), \tag{1.3}$$

where $0 \leqslant \text{arc cos } x \leqslant \pi$.

We recall that

$$e^{i\theta} = \cos \theta + i \sin \theta$$

and

$$e^{in\theta} = (\cos \theta + i \sin \theta)^n = \cos n\theta + i \sin n\theta. \tag{1.4}$$

By the binomial expansion

$$(\cos \theta + i \sin \theta)^n = \cos^n \theta + \binom{n}{1} \cos^{n-1} \theta(i \sin \theta)$$

$$+ \binom{n}{2} \cos^{n-2} \theta(i^2 \sin^2 \theta) + \cdots + \binom{n}{n}(i \sin \theta)^n.$$

Equating the real parts of the last equation of (1.4), we obtain

$$\cos n\theta = \cos^n \theta - \binom{n}{2}\cos^{n-2}\theta \sin^2 \theta + \binom{n}{4}\cos^{n-4}\theta \sin^4\theta + \cdots$$

$$+ (-1)^{[n/2]} \binom{n}{2[n/2]}\cos^{n-2[n/2]}\theta \, \sin^{2[n/2]}\theta .\dagger \tag{1.5}$$

Note that only even powers of sin θ occur in (1.5). We therefore make the substitution $\sin^2 \theta = 1 - \cos^2 \theta$ in (1.5) and obtain

$$\cos n\theta = \sum_{q=0}^{[n/2]} (-1)^q \binom{n}{2q} \cos^{n-2q}\theta \left(\sum_{k=0}^{q} (-1)^k \binom{q}{k} \cos^{2k}\theta \right). \tag{1.6}$$

The right-hand side of (1.6) is a polynomial in $x = \cos \theta$, and so our function $T_n(x)$, defined in (1.3), is a polynomial. We proceed to determine its coefficients.

The right-hand side of (1.6) is a "triangular" sum; namely, if we write

$$A_q = (-1)^q \binom{n}{2q} \cos^{n-2q}\theta, \qquad q = 0, \ldots, \left[\frac{n}{2}\right]$$

---

† $[y]$ means the greatest integer not exceeding y; e.g., if n is even $[n/2] = n/2$, whereas if n is odd $[n/2] = (n-1)/2$.

and

$$B_{k,q} = (-1)^k \binom{q}{k} \cos^{2k} \theta, \qquad k = 0, 1, \ldots, q,$$

then

$$\begin{aligned}
\cos n\theta = {}& A_0 B_{0,0} \\
& + A_1 B_{0,1} + A_1 B_{1,1} \\
& + A_2 B_{0,2} + A_2 B_{1,2} + A_2 B_{2,2} \\
& + \\
& \vdots \\
& + A_{[n/2]} B_{0,[n/2]} + \cdots + A_{[n/2]} B_{[n/2],[n/2]} .
\end{aligned} \tag{1.7}$$

Let us add up the right-hand side of (1.7) by stripping off successive diagonals. We then obtain

$$\begin{aligned}
\cos n\theta = {}& (A_0 B_{0,0} + A_1 B_{1,1} + \cdots + A_{[n/2]} B_{[n/2],[n/2]}) \\
& + (A_1 B_{0,1} + A_2 B_{1,2} + \cdots + A_{[n/2]} B_{[n/2]-1,[n/2]}) \\
& + \\
& \vdots \\
& + (A_{[n/2]-1} B_{0,[n/2]-1} + A_{[n/2]} B_{1,[n/2]}) \\
& + A_{[n/2]} B_{0,[n/2]} ;
\end{aligned}$$

or, by replacing the A_q and $B_{k,q}$ with what they stand for

$$\cos n\theta = \sum_{k=0}^{[n/2]} \left((-1)^k \sum_{j=k}^{[n/2]} \binom{n}{2j}\binom{j}{k} \right) \cos^{n-2k} \theta. \tag{1.8}$$

Equation 1.8 reveals that $T_n(x)$ is a polynomial of degree n.
   If we write

$$T_n(x) = t_0^{(n)} + t_1^{(n)} x + \cdots + t_n^{(n)} x^n, \tag{1.9}$$

we deduce from (1.8) that

$$t_{n-(2k+1)}^{(n)} = 0, \qquad k = 0, \ldots, \left[\frac{n-1}{2}\right],$$

$$t_{n-2k}^{(n)} = (-1)^k \sum_{j=k}^{[n/2]} \binom{n}{2j}\binom{j}{k}, \qquad k = 0, \ldots, \left[\frac{n}{2}\right]. \tag{1.10}$$

Thus $T_n(x)$, which was defined in (1.2) by its values in I, turns out to be a polynomial of degree n, hence is defined for all x (indeed for all complex numbers x). $T_n(x)$ is called the *Chebyshev polynomial* of degree n. For each nonnegative integer n the Chebyshev polynomial of degree n is given explicitly by formulas (1.9) and (1.10). Let us list the first few Chebyshev polynomials obtained from (1.9) and (1.10):

$$T_0(x) = 1; \qquad T_1(x) = x; \qquad T_2(x) = 2x^2 - 1;$$
$$T_3(x) = 4x^3 - 3x; \qquad T_4(x) = 8x^4 - 8x^2 + 1; \qquad (1.11)$$
$$T_5(x) = 16x^5 - 20x^3 + 5x.$$

$T_0, T_1, \ldots, T_5$ are graphed in Figure 1.1.

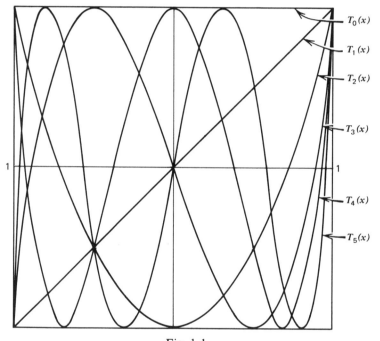

$T_0(x)$

$T_1(x)$

$T_2(x)$

$T_3(x)$

$T_4(x)$

$T_5(x)$

Fig. 1.1

The sequence of polynomials $\{T_n(x)\}_{n=0}^{\infty}$ is named after the Russian mathematician P. L. Chebyshev (1821–1894) who first studied them. The collected works of this eminent savant are available in Russian and French (Tchebychef [1].) This book is devoted to the study of various properties of these polynomials. In using the notation $T_n(x)$ for the Chebyshev polynomial of degree n we are following traditional usage derived from another transliteration of the name Chebyshev in the form Tchebycheff or related forms.

EXERCISES 1.1

1.1.1. Show that

$$T_n(x) = \tfrac{1}{2}[(x + \sqrt{x^2 - 1})^n + (x - \sqrt{x^2 - 1})^n]. \tag{1.12}$$

Hint. Suppose that $x \in I$ and recall that $\cos n\theta = (e^{in\theta} + e^{-in\theta})/2$.

1.1.2. If $x \geqslant 1$, prove that

$$T_n(x) = \cosh nt,$$

where $x = \cosh t, t \geqslant 0$.

As a consequence of (1.2) any trigonometric identities involving $\cos n\theta$ can be carried over immediately to identities involving the Chebyshev polynomials. The following exercises illustrate this theme. In the absence of other instructions the reader should attempt to verify an exercise.

1.1.3. If m, n are nonnegative integers, then

$$T_m(x)T_n(x) = \tfrac{1}{2}(T_{m+n}(x) + T_{|m-n|}(x)).$$

1.1.4. $\int T_n(x)\,dx = \tfrac{1}{2}\left(\dfrac{T_{n+1}(x)}{n+1} - \dfrac{T_{n-1}(x)}{n-1} \right) + C, \qquad n \geqslant 2.$

1.1.5. $(T_{m+n}(x) - 1)(T_{|m-n|}(x) - 1) = (T_m(x) - T_n(x))^2.$

1.1.6. $T_m(T_n(x)) = T_{mn}(x)$ for all nonnegative integers m and n.

1.2. Some Simple Properties

Now that we have defined the Chebyshev polynomials and written down an explicit formula for them we proceed to investigate some of their simpler properties.

Formula 1.10 reveals that for even n only even powers of x occur in $T_n(x)$, whereas for odd n only odd powers of x occur. Thus for all nonnegative integers n

$$T_n(-x) = (-1)^n T_n(x); \tag{1.13}$$

i.e., $T_n(x)$ is an even function for even n and an odd function for odd n.

We also observe from (1.10) that the nonzero coefficients of $T_n(x)$ are integers that alternate in sign, the leading coefficient, $t_n^{(n)} = t_n$,† being positive. Indeed, if $n > 0$, $t_n^{(n)}$ has a particularly simple form for

$$t_n = \sum_{j=0}^{[n/2]} \binom{n}{2j} = \tfrac{1}{2}\{(1 + 1)^n + (1 - 1)^n\} = 2^{n-1}. \tag{1.14}$$

† We have written the coefficient of x^k in $T_n(x)$, $t_k^{(n)}$, with the superscript (n), to exhibit the fact that the Chebyshev polynomials of different degree have independent sets of coefficients, but when the degree of the polynomial is fixed in a discussion and no confusion results we simply drop the superscript.

For some purposes the polynomial with leading coefficient 1,

$$\tilde{T}_n(x) = 2^{1-n} T_n(x) = x^n + 2^{1-n} \sum_{j=0}^{n-1} t_j x^j, \qquad n > 0,$$

$$(\tilde{T}_0(x) = 1),$$

(1.15)

is useful.

We turn next to the significant *points* of $T_n(x)$. These are the zeros and extrema of T_n. The zeros are, of course, simply the values of x for which $T_n(x) = 0$. Since $T_n(x) = \cos n\theta$ and $\cos n\theta_j = 0$ for

$$\theta_j = \theta_j^{(n)} = \frac{(2j-1)}{n} \frac{\pi}{2}, \qquad j = 1, \ldots, n,$$

(1.16)

we see that the points

$$\xi_j = \xi_j^{(n)} = \cos \theta_j^{(n)} = \cos \frac{(2j-1)}{n} \frac{\pi}{2}, \qquad j = 1, \ldots, n$$

(1.17)

are all distinct, lie in I, and satisfy

$$T_n(\xi_j) = 0, \qquad j = 1, \ldots, n.$$

(Once again we omit superscripts when n is fixed.) Since $T_n(x)$ is of degree n, it has exactly n zeros and so the numbers $\xi_j^{(n)}, j = 1, \ldots, n$, defined in (1.17), are all the zeros of $T_n(x)$.

It is clear from (1.2) that if $x \in I$

$$-1 \leqslant T_n(x) \leqslant 1.$$

The points of I at which $|T_n(x)| = 1$ we call the *extrema* of $T_n(x)$. We know that $\cos k\pi = (-1)^k$ for any integer k; hence if

$$\phi_k = \phi_k^{(n)} = \frac{k\pi}{n}, \qquad k = 0, 1, \ldots, n,$$

(1.18)

the points

$$\eta_k = \eta_k^{(n)} = \cos \phi_k^{(n)} = \cos \frac{k\pi}{n}, \qquad k = 0, 1, \ldots, n,$$

(1.19)

are all distinct, lie in I, and satisfy

$$T_n(\eta_k) = (-1)^k, \qquad k = 0, \ldots, n.$$

(1.20)

The points $\eta_0, \ldots, \eta_n$ are the extrema of $T_n(x)$. It is clear that, since $|T_n(x)| \leqslant 1$ for $x \in I$, the points $\eta_1, \ldots, \eta_{n-1}$ which lie in the interior of I are *relative extrema* of $T_n(x)$ so that

$$T_n'(\eta_k) = 0, \qquad k = 1, \ldots, n-1.$$

(1.21)

Since T'_n is a polynomial of degree $n - 1$, all its zeros are $\eta_1, \ldots, \eta_{n-1}$ and so the points $\eta_0 = 1$ and $\eta_n = -1$ are not relative extrema of $T_n(x)$.

On differentiating $T_n(x) = \cos n\theta$ with respect to x we obtain

$$T'_n(x) = \left(\frac{d}{d\theta} \cos n\theta\right) \frac{d\theta}{dx} = \frac{-n \sin n\theta}{-\sin \theta} = n \frac{\sin n\theta}{\sin \theta}, \qquad x = \cos \theta. \qquad (1.22)$$

The polynomial of degree $n - 1$

$$U_{n-1}(x) = \frac{1}{n} T'_n(x) = \frac{\sin n\theta}{\sin \theta}, \qquad (x = \cos \theta), \qquad (1.23)$$

is called *the Chebyshev polynomial of the second kind*. Its zeros are $\eta_1, \ldots, \eta_{n-1}$ and its explicit form is easily obtained by differentiating (1.9) and dividing by n.

Because we refer frequently to the zeros and extrema of $T_n(x)$, the reader should note that both $\xi_1, \ldots, \xi_n$ and $\eta_0, \ldots, \eta_n$ move from right to left in I with increasing index.

EXERCISES 1.2

1.2.1. Show that if $n = 2m$

$$T_n(x) = \tau_m(x^2),$$

and if $n = 2m + 1$

$$T_n(x) = x \, \tau_m(x^2),$$

where

$$\tau_m(t) = \sum_{j=0}^{m} (-1)^j \binom{n}{2j} t^{m-j}(1 - t)^j.$$

1.2.2. Find all the solutions of $x = T_n(x)$, $\qquad n = 2, 3, \ldots$.

1.2.3. Verify that

$$T'_n(\xi_j) = (-1)^{j-1} \frac{n}{\sin(2j - 1)(\pi/2n)}, \qquad j = 1, \ldots, n,$$

$$= (-1)^{j-1} \frac{n}{\sqrt{1 - \xi_j^2}}.$$

1.2.4. Show that

$$|T'_n(x)| \leqslant n^2, \qquad x \in I, \qquad (1.24)$$

with equality holding only if $x = \pm 1$.

Hint. Use the representation (1.22) and mathematical induction on n.

1.2.5. Show that if $n = 2rk + m$; $r \geqslant 0, -(k-1) \leqslant m \leqslant k$,

$$T_n(\xi_j^{(k)}) = (-1)^r \, T_m(\xi_j^{(k)}).$$

1.2.6. Show that any polynomial $p(x) = a_0 + a_1 x + \cdots + a_n x^n$ can be written $p(x) = b_0 + b_1 T_1(x) + \cdots + b_n T_n(x)$ and $b_n = 2^{-(n-1)} a_n$.

1.2.7. Show that

$$T_n(\xi_j^{(n+1)}) = (-1)^{j-1} \sqrt{1 - (\xi_j^{(n+1)})^2}, \qquad n = 1, 2, \ldots; \qquad j = 1, 2, \\ \ldots, n+1,$$

and

$$T_{n+1}(\xi_j^{(n)}) = (-1)^j \sqrt{1 - (\xi_j^{(n)})^2}, \qquad n = 0, 1, 2, \ldots, \qquad j = 1, 2, \ldots, n.$$

1.2.8. Show that

$$T_n''(\xi_j) = (-1)^{j-1} n \frac{\xi_j}{(1 - \xi_j^2)^{3/2}}.$$

1.2.9. Show that

$$\sum_{j=1}^{n} \xi_j^{(n)} = 0.$$

1.2.10. Show that

$$T_n'(x) = \sum_{i=1}^{n} \frac{T_n(x)}{x - \xi_i}.$$

1.2.11. Show that for $n \geqslant 2$ the only solution of $T_n(x) = T_n'(x)$ that satisfies $x > 1$ lies in $(n, n + 1/2n)$.

1.2.12. Suppose $p(x) = (T_n(x))^{2k+1}$ in Exercise 1.2.6 and that

$$(T_n(x))^{2k+1} = B_0 + B_1 T_1(x) + \cdots + B_{n(2k+1)} T_{n(2k+1)}(x).$$

Show that $B_0 = B_1 = \cdots = B_{n-1} = 0$.

Hint. Use the fact that Exercise 1.1.3 implies that $T_n^{2k} = ((1 + T_{2n})/2)^k$, and then use Exercise 1.1.3 repeatedly.

1.2.13. Show that

$$\tfrac{1}{2} + T_2(x) + T_4(x) + \cdots + T_{2k}(x) = \frac{U_{2k}(x)}{2}, \qquad k = 0, 1, \ldots.$$

Hint. $\sin A \cos B = \tfrac{1}{2} [\sin (A - B) + \sin (A + B)]$.

1.2.14. If $U_n(x) = u_0 + u_1 x + \cdots + u_n x^n$, then $u_n = 2^n$.

1.2.15. (a) $U_n(x) - U_{n-2}(x) = 2T_n(x)$,
(b) $T_n(x) = U_n(x) - x\,U_{n-1}(x)$,
(c) $\dfrac{U_{2k+1}(x)}{2} = T_1(x) + T_3(x) + \cdots + T_{2k+1}(x)$, $\quad k = 0, 1, \ldots,$
(d) $x\,U_{2k-1}(x) = 1 + 2T_2(x) + \cdots + 2T_{2k-2}(x) + T_{2k}(x)$,
$k = 1, 2, \ldots$.
(e) $U_{nm-1}(x) = U_{m-1}(T_n(x))\,U_{n-1}(x)$.
(f) $U_n(x) = \sum_{j=0}^{n} x^j\,T_{n-j}(x)$.

1.2.16. Let $s_k^{(n)} = t_0^{(n)} + t_1^{(n)} + \cdots + t_k^{(n)}$, $\quad k = 0, 1, \ldots, n$. If $q(x) = s_0 + s_1 x + \cdots + s_n x^n$, show that $p(x) = (1 - x)\,q(x) = T_n(x) - x^{n+1}$.

1.2.17. Show that $p(x) = T_n(x) - x^{n+1}$ has at least $[n/2]$ distinct zeros in $(0, 1]$.

1.2.18. Show that $p(x) = T_n(x) - x^{n+1}$ has $[n/2]$ positive zeros other than $x = 1$.

Hint. According to Descartes' rule of signs, if $p(x) = a_0 + a_1 x + \cdots + a_{n+1} x^{n+1}$ has N positive zeros and there are W changes of sign in the sequence $a_0, a_1, \ldots, a_{n+1}$, then $W - N$ is either zero or a positive even number; but, in view of (1.10), $W - [n/2] + 1$.

1.2.19. Show that the numbers $s_{n-2j}^{(n)}$, $j = 0, \ldots, [n/2]$, defined in Exercise 1.2.16, alternate in sign.

Hint. Use Descartes' rule of signs again, this time starting with information about the number of zeros of $q(x)$.

1.3. Polynomial Interpolation at the Zeros and Extrema

The zeros and extrema of the Chebyshev polynomials play an important role in the theory of polynomial interpolation. The setting is the following.

Suppose $f(x)$ is a continuous function defined on I, which we wish to approximate by a polynomial of degree at most k. As a measure of how good an approximation of $f(x)$ is provided by a given $p \in \mathscr{P}_k$ we adopt the *uniform norm*

$$\|f - p\| = \max_{1 \leqslant x \leqslant 1} |f(x) - p(x)|;$$

i.e., the measure of approximation is the greatest distance between $f(x)$ and $p(x)$ as x runs through I. A rather natural way to seek polynomial approximations to $f(x)$ is to sample $f(x)$ at distinct points of I, $x_1, \ldots, x_n$, and try to find a polynomial that takes on the same values as $f(x)$ at $x_1, \ldots, x_n$. Such a polynomial is said to *interpolate* $f(x)$ at the *nodes* $x_1, \ldots, x_n$. As a matter of fact, we shall now show that, given distinct points of I, $x_1, \ldots, x_n$, it is easy to construct a unique $p \in \mathscr{P}_{n-1}$ that interpolates $f(x)$ at $x_1, \ldots, x_n$.

We wish to construct a polynomial that passes through the points $(x_1, f(x_1))$, $(x_2, f(x_2)), \ldots, (x_n, f(x_n))$.

Let us put

$$l_{j,n}(x) = l_j(x) = \frac{(x - x_1)(x - x_2) \cdots (x - x_{j-1})(x - x_{j+1}) \cdots (x - x_n)}{(x_j - x_1)(x_j - x_2) \cdots (x_j - x_{j-1})(x_j - x_{j+1}) \cdots (x_j - x_n)},$$

$$j = 1, \ldots, n, \quad n \geqslant 1. \tag{1.25}$$

$l_j(x)$ is a polynomial of degree $n - 1$ that satisfies

$$l_j(x_i) = \begin{cases} 0, & j \neq i, \\ 1, & j = i, \end{cases} \qquad i, j = 1, \ldots, n, \tag{1.26}$$

as is readily evident from (1.25); $l_{1,1}$ is identically 1 and $l_1(x), \ldots, l_n(x)$ are called the *fundamental polynomials* for interpolation at $x_1, \ldots, x_n$.

$$L_{n-1}(x) = f(x_1) \, l_1(x) + f(x_2) \, l_2(x) + \cdots + f(x_n) \, l_n(x) \tag{1.27}$$

is a polynomial of degree at most $n - 1$ that passes through the points in question. Moreover, if $p \in \mathscr{P}_{n-1}$ and p interpolates f at $x_1, \ldots, x_n$, then $p = L_{n-1}$, for if $p(x_j) = f(x_j), j = 1, \ldots, n$ then

$$L_{n-1}(x_j) - p(x_j) = 0, \qquad j = 1, \ldots, n,$$

and the polynomial $L_{n-1} - p \in \mathscr{P}_{n-1}$ has n distinct zeros which means that $L_{n-1} = p$. Thus $L_{n-1}(x)$, as defined in (1.27), is the unique member of $\mathscr{P}_{n-1}$ that interpolates $f(x)$ at $x_1, \ldots, x_n$. This unique interpolating polynomial, when written in the form (1.27), is called the Lagrange interpolating polynomial [to $f(x)$ at $x_1, \ldots, x_n$].

If we start with an infinite triangular array of nodes,

$$X: \qquad \begin{array}{l} x_1^{(1)} \\ x_1^{(2)}, x_2^{(2)} \\ \vdots \\ x_1^{(n)}, x_2^{(n)}, \ldots, x_n^{(n)} \\ \vdots \end{array} \tag{1.28}$$

where for $n = 1, 2, \ldots$ each $x_j^{(n)} \in I, j = 1, \ldots, n$, the rows of X determine a sequence of interpolating polynomials

$$\{L_k\}_{k=0}^{\infty}, \tag{1.29}$$

the polynomial $L_{n-1} \in \mathscr{P}_{n-1}$ being the unique interpolating polynomial determined by the nth row. The notation L_k for a member of the sequence (1.29) is shorthand for $L_k(f, X; x)$, in which the subscript indicates an element of $\mathscr{P}_k$ obtained by interpolating the first argument $f(x)$ at the entries in the $(k + 1)$st row of the second argument X.

Given X, the sequence (1.29) provides us with approximating polynomials to $f(x)$ on I. How good an approximation these polynomials are is, as we assumed, measured by the numbers

$$M_k = \|f - L_k\| = \max_{-1 \leqslant x \leqslant 1} |f(x) - L_k(x)|, \qquad k = 0, 1, 2, \ldots.$$

We wish to compare M_k with the best approximation possible by means of $p \in \mathscr{P}_k$. It is known (cf. Rivlin [1]) that there is a $p^* \in \mathscr{P}_k$ that gives this best approximation; i.e., given $f(x)$,

$$\|f - p^*\| \leqslant \|f - p\|$$

for all $p \in \mathscr{P}_k$. We put

$$E_k(f) = \|f - p^*\|.$$

We can now prove a result comparing M_k with E_k.

THEOREM 1.1

$$M_k \leqslant E_k(1 + \max_{-1 \leqslant x \leqslant 1} \sum_{j=1}^{k+1} |l_j(x)|), \qquad k = 0, 1, \ldots. \tag{1.30}$$

Proof. After subtracting and adding p^* to $f - L_k$, we obtain

$$|f(x) - L_k(f, X; x)| \leqslant |f(x) - p^*(x)| + |p^*(x) - L_k(f, X; x)|. \tag{1.31}$$

Now, if $p \in \mathscr{P}_k$,

$$L_k(p, X; x) = p(x),$$

since $L_k(p, X; x)$ is the unique interpolating polynomial to $p(x)$ in the $(k+1)$st row of X and p surely interpolates itself. Hence, in particular,

$$p^*(x) = L_k(p^*, X; x),$$

and in view of (1.27)

$$p^*(x) - L_k(f, X; x) = L_k(p^*, X; x) - L_k(f, X; x)$$
$$= L_k(p^* - f, X; x).$$

From (1.31) we obtain

$$|f(x) - L_k(f, X; x)| \leqslant E_k + |L_k(p^* - f, X; x)|, \tag{1.32}$$

but, in general, if $g(x)$ is continuous on I,

$$|L_k(g, X; x)| \leqslant |g(x_1^{(k+1)}) l_{1, k+1}(X; x)| + \cdots + |g(x_{k+1}^{(k+1)}) l_{k+1, k+1}(X; x)|$$

(where $l_{j,k+1}(X;x)$ is a full notation for (1.25) with the $(k+1)$st row of X as the nodes), and so

$$|L_k(g, X; x)| \leqslant \max_{-1 \leqslant x \leqslant 1} |g(x)| \max_{-1 \leqslant x \leqslant 1} \sum_{j=1}^{k+1} |l_j(x)|. \tag{1.33}$$

If we apply (1.33) with $g = p^* - f$ and note that

$$E_k = \max_{-1 \leqslant x \leqslant 1} |p^*(x) - f(x)|,$$

we obtain from (1.32)

$$|f(x) - L_k(f, X; x)| \leqslant E_k \left(1 + \max_{-1 \leqslant x \leqslant 1} \sum_{j=1}^{k+1} |l_j(x)|\right). \tag{1.34}$$

The theorem now follows by choosing x on the left-hand side of (1.34) so that $|f(x) - L_k(f, X; x)| = \|f - L_k\|$. ∎

The function

$$\lambda_{k+1}(X; x) = \sum_{j=1}^{k+1} |l_{j,k+1}(X; x)|, \tag{1.35}$$

which appears in (1.30), is called the *Lebesgue function* of order $k + 1$ of X. Note that it does not depend on $f(x)$. The quantity

$$\Lambda_{k+1}(X) = \max_{-1 \leqslant x \leqslant 1} \lambda_{k+1}(X; x)$$

is called the *Lebesgue constant* of order $k + 1$ of X; (1.30) may now be written concisely as

$$M_k \leqslant E_k(1 + \Lambda_{k+1}), \qquad k = 0, 1, \ldots, \tag{1.36}$$

the various dependencies on f and X being tacitly understood as usual.

Since E_k depends on f and k, but not on X, the effect on M_k of X, insofar as (1.36) is informative, comes from the Lebesgue constant Λ_{k+1}. Formula tells us that the smaller $\Lambda_k(X)$, the better the sequence of Lagrange interpolating polynomials at the nodes of X as uniform approximations of f. It is a fact that there is an array of nodes X^* such that

$$\Lambda_k(X^*) \leqslant \Lambda_k(X), \qquad k = 1, 2, 3, \ldots,$$

for any array of nodes X, but the determination of a "best" array of nodes X^* and the corresponding Lebesgue constants is an open question. The point of this digression on the topic of polynomial interpolation is that the zeros of the Chebyshev polynomials provide an array of nodes with "small" Lebesgue constants. We proceed now toward making this assertion more precise.

It has been shown by Erdös [1] that there exists a positive constant, c, such that

$$\Lambda_k(X) > \frac{2}{\pi} \log k - c, \qquad k = 1, 2, \ldots, \tag{1.37}$$

for any X. This shows, incidentally, that $\Lambda_k(X) \to \infty$ as $k \to \infty$, a fact with the startling consequence (Faber [1]) that, given X, there exists a function, $f(x)$, continuous on I such that $\{L_k(f, X; x\}$ does *not* converge uniformly to $f(x)$. (We cannot pause to prove this here, nor shall we prove (1.37) or even similar, less sharp, results. The reader interested in these proofs is referred to the original papers or should consult Natanson [1].) Thus our original hope of approximating *all* continuous functions by interpolating polynomials, using a fixed X, turns out to be illusory. We shall, however, show that

$$\Lambda_k(T) \leqslant \frac{2}{\pi} \log k + 1, \qquad k = 1, 2, \ldots, \tag{1.38}$$

where T is the array whose kth row is $\xi_1^{(k)}, \ldots, \xi_k^{(k)}$, i.e., the zeros of $T_k(x)$. In view of (1.37) and (1.36), although T may not be the best array of nodes for interpolation, it is an excellent choice.

Let us see what $L_{n-1}(f, T; x)$ looks like. We remark first that if we put

$$\omega(x) = (x - x_1)(x - x_2) \cdots (x - x_n),$$

then $l_j(x)$, as defined in (1.25), can be written as

$$l_j(x) = \frac{\omega(x)}{(x - x_j)\omega'(x_j)}, \qquad j = 1, \ldots, n, \tag{1.39}$$

so that (1.27) becomes

$$L_{n-1}(x) = \omega(x) \sum_{j=1}^{n} \frac{f(x_j)}{(x - x_j)\,\omega'(x_j)}. \tag{1.40}$$

When T is the array of nodes, $\omega(x) = \widetilde{T}_n(x)$ and so

$$\begin{aligned}
L_{n-1}(f, T; x) &= T_n(x) \sum_{j=1}^{n} \frac{f(\xi_j^{(n)})}{(x - \xi_j^{(n)})\,T_n'(\xi_j^{(n)})} \\
&= \frac{T_n(x)}{n} \sum_{j=1}^{n} (-1)^{j-1}\, \frac{f(\xi_j)\, \sin\,[(2j-1)\pi]/2n}{(x - \xi_j)} \\
&= \frac{T_n(x)}{n} \sum_{j=1}^{n} (-1)^{j-1}\, \frac{f(\xi_j)(1 - \xi_j^2)^{1/2}}{(x - \xi_j)},
\end{aligned} \tag{1.41}$$

where we have used Exercise 1.2.3.

If we use the trigonometric form, (1.41) becomes

$$L_{n-1}(f, T; \cos \theta) = \frac{\cos n\theta}{n} \sum_{j=1}^{n} (-1)^{j-1} \frac{f(\cos \theta_j)}{\cos \theta - \cos \theta_j} \sin \theta_j, \tag{1.42}$$

and the Lebesgue function may be written

$$\lambda_n(T; x) = \lambda_n(\cos \theta) = \frac{|\cos n\theta|}{n} \sum_{j=1}^{n} \frac{\sin \theta_j}{|\cos \theta - \cos \theta_j|}. \tag{1.43}$$

To establish (1.38) we show first, following Ehlich and Zeller [2], that $\Lambda_n(T) = \lambda_n(T; 1)$ for $n \geqslant 2$ (that $\lambda_1(T, 1) = \Lambda_1(T) = 1$ is a trivial observation). To this end we need some information about trigonometric polynomials. A trigonometric polynomial of degree k is a function

$$t(\theta) = \sum_{j=0}^{k} (a_j \cos j\theta + b_j \sin j\theta),$$

with $a_k^2 + b_k^2 > 0$. We suppose that, unless otherwise stated, the coefficients $a_0, \ldots, a_k; b_0, \ldots, b_k$ are real numbers. The set of trigonometric polynomials of degree at most n is denoted by $\mathscr{T}_n$. (The zero polynomial is arbitrarily assigned the degree -1.) A nonzero trigonometric polynomial of degree k has at most $2k$ zeros in the interval $[0, 2\pi)$, where multiple zeros are counted as distinct; i.e., a zero of multiplicity m is counted as m zeros. We leave this fact as an exercise (Exercise 1.3.13).

Let us put [cf. (1.16)]

$$\theta_j = (2j - 1)\frac{\pi}{2n}, \qquad j = 0, \pm 1, \pm 2, \ldots,$$

and

$$d_k(\theta) = \frac{1}{2n} \frac{\sin n(\theta - \theta_k)}{\tan \frac{1}{2}(\theta - \theta_k)}, \qquad k = 0, \pm 1, \pm 2, \ldots; \tag{1.44}$$

then $d_k(\theta) \in \mathscr{T}_n$. To verify this we observe that

$$d_k(\theta) = \frac{1}{2n} \frac{\sin 2n(\theta - \theta_k)/2}{\sin (\theta - \theta_k)/2} \cos \frac{\theta - \theta_k}{2}, \tag{1.45}$$

which implies, in view of the trigonometric form of Exercise 1.2.15d, that

$$d_k(\theta) = \frac{1}{2n} \left[1 + 2 \sum_{j=1}^{n-1} \cos j(\theta - \theta_k) + \cos n(\theta - \theta_k) \right]. \tag{1.46}$$

Furthermore, for $j = 1, \ldots, 2n; k = 1, \ldots, 2n$

$$d_k(\theta_j) = \begin{cases} 0, & j \neq k, \\ 1, & j = k. \end{cases} \tag{1.47}$$

Thus the functions $d_k(\theta)$, $k = 1, \ldots, 2n$ are fundamental polynomials for interpolation by trigonometric polynomials of degree at most n at $\theta_1, \ldots, \theta_{2n}$; i.e.,

$$t(\theta) = \sum_{k=1}^{2n} y_k d_k(\theta) \tag{1.48}$$

satisfies

$$t(\theta_j) = y_j, \qquad j = 1, \ldots, 2n. \tag{1.49}$$

The trigonometric Lebesgue function

$$\delta_n(\theta) = \sum_{k=1}^{2n} |d_k(\theta)|$$

has the property that

$$\delta_n\left(\theta + \frac{\pi}{n}\right) = \delta_n(\theta). \tag{1.50}$$

This follows from the observations that $d_k(\theta + \pi/n) = d_{k-1}(\theta)$, and $d_0(\theta) = d_{2n}(\theta)$. As a consequence of (1.50)

$$\Delta_n = \max_{0 \leqslant \theta \leqslant 2\pi} |\delta_n(\theta)| = \max_{-\pi/2n \leqslant \theta \leqslant \pi/2n} |\delta_n(\theta)|.$$

Now for each $k = 1, \ldots, 2n, d_k(\theta_j) = 0, j \neq k, j = 1, \ldots, 2n$, and, in addition, $d_k'(\theta_{k+n}) = d_k'(\theta_{k-n}) = 0$. Note that either $1 \leqslant k + n \leqslant 2n$ or $1 \leqslant k - n \leqslant 2n$, so that for each $d_k(\theta)$ we have accounted for $2n - 2$ simple zeros and one double zero, i.e., for all $2n$ of its zeros in $[0, 2\pi)$. Hence, for $k = 1, \ldots, 2n, d_k(\theta) \neq 0$ in $(-\pi/2n, \pi/2n)$ and $\delta_n(\theta)$ coincides with a trigonometric polynomial

$$t(\theta) = \sum_{k=1}^{2n} \epsilon_k d_k(\theta)$$

in the interval $[-\pi/2n, \pi/2n]$, where $\epsilon_k = \pm 1$, the sign being chosen so that $\epsilon_k d_k(\theta) > 0$ for $-\pi/2n < \theta < \pi/2n$. Therefore ϵ_k has the same sign as

$$d_k(0) = \frac{1}{2n} \frac{\sin n\theta_k}{\tan(\theta_k/2)} = \frac{(-1)^{k-1}}{2n} \frac{1}{\tan^{(2k-1)}(\pi/4n)},$$

i.e.,

$$\epsilon_k = (-1)^{k-1}, \qquad k = 1, \ldots, n,$$
$$\epsilon_k = (-1)^k, \qquad k = n+1, \ldots, 2n,$$

and so,

$$t(\theta) = \sum_{k=1}^{n} (-1)^{k-1} d_k(\theta) - \sum_{k=n+1}^{2n} (-1)^{k-1} d_k(\theta). \qquad (1.51)$$

A simple computation next reveals that $d_k(-\theta) = d_{2n-k+1}(\theta)$ and consequently $t(\theta) = t(-\theta)$, so that t is an even function.

Let

$$\Delta_n = \max_{-\pi/2n \le \theta \le \pi/2n} t(\theta) = t(\bar\theta).$$

We claim that $\bar\theta = 0$.

First note that $\bar\theta \neq \pm\pi/(2n)$, for if $\bar\theta = \pm\pi/(2n)$ then $t(\bar\theta) = 1$ in view of (1.51) and (1.47) and $\Delta_n = 1$, but

$$s(\theta) = \sum_{k=1}^{2n} d_k(\theta)$$

satisfies $s(\theta_j) = 1, j = 1, \ldots, 2n$, hence $1 - s(\theta) = u(\theta)$, where $u \in \mathcal{T}_n$ either has simple zeros at $\theta_1, \ldots, \theta_{2n}$ or is identically zero. In the former case $u(\theta)$ changes sign at the θ_j. In particular, then, there exists θ^* close to θ_1 so that $u(\theta^*) < 0$ and $s(\theta^*) = 1 - u(\theta^*) > 1$ and

$$1 = \Delta_n \ge \delta_n(\theta^*) \ge |s(\theta^*)| > 1$$

gives a contradiction. If $u = 0$, then $s = 1$ and $\delta_n(\theta) \equiv 1$, which implies that $t(\theta) \equiv 1$. However, since $n \ge 2$, we have $t(\theta_2) = -1$ in view of (1.51) and (1.47), again giving a contradiction.

Suppose that $0 < |\bar\theta| < \pi/2n$; then, because of the evenness of t, $t(\bar\theta) = t(-\bar\theta)$, and by Rolle's theorem t' has a zero between $-\bar\theta$ and $\bar\theta$, in addition to $t'(\bar\theta) = t'(-\bar\theta) = 0$, for a total of at least three distinct zeros in $(-\pi/(2n), \pi/(2n))$. Also, from our previous observation that $d_k(\theta + \pi/n) = d_{k-1}(\theta)$ it follows that $d_k(\theta + \pi) = d_{k-n}(\theta) = d_{k+n}(\theta)$; hence $t(\theta + \pi) = (-1)^{n+1} t(\theta)$. Therefore $t'(\theta)$ also has at least three distinct zeros in (θ_n, θ_{n+1}). Since $t(\theta_k) = (-1)^{k-1}, k = 1, \ldots, n$, t has at least $n - 1$ distinct zeros in (θ_1, θ_n) and by Rolle's theorem t' has at least $n - 2$ zeros in (θ_1, θ_n). Similarly, $t(\theta_k) = (-1)^k, k = n + 1, \ldots, 2n$ and t' has at least $n - 2$ zeros in $(\theta_{n+1}, \theta_{2n})$. Thus $t' \in \mathcal{T}_n$ has at least $2n - 4 + 6 = 2n + 2$ zeros, hence is identically zero, and t is a constant, but $t(\theta_1) = 1$ and $t(\theta_2) = -1(n \ge 2)$, a contradiction. We have proved that $\Delta_n = \delta_n(0)$. Since

$$(-1)^{k-1} d_k(0) = \frac{1}{2n} \cot\frac{(2k-1)\pi}{4n},$$

we obtain

$$\Delta_n = \frac{1}{n} \sum_{k=1}^{n} \cot \frac{(2k-1)\pi}{4n}. \tag{1.52}$$

Observe that

$$\frac{\pi}{2} \Delta_n = \frac{\pi}{2n} \sum_{k=1}^{n} \left(\cot \frac{(2k-1)\pi}{4n} - \frac{4n}{(2k-1)\pi} \right) + 2 \sum_{k=1}^{n} \frac{1}{2k-1} \tag{1.53}$$

$$= a_n + 2 \sum_{k=1}^{n} \frac{1}{2k-1};$$

hence

$$\frac{\pi}{2} \Delta_n - \log n = a_n + 2 \sum_{k=1}^{n} \frac{1}{2k-1} - \log n.$$

The a_n form a sequence of Riemann sums of the integral

$$\int_0^{\pi/2} \left(\cot x - \frac{1}{x} \right) dx = \log \frac{2}{\pi};$$

hence

$$\lim_{n \to \infty} a_n = \log \frac{2}{\pi},$$

whereas

$$? \sum_{k=1}^{n} \frac{1}{2k-1} - \log n = 2 \left(\sum_{k=1}^{2n} \frac{1}{k} - \log 2n \right) - \left(\sum_{k=1}^{n} \frac{1}{k} - \log n \right) + \log 4.$$

Since we know that

$$\gamma = \lim_{m \to \infty} \left(\sum_{j=1}^{m} \frac{1}{j} - \log m \right) = 0.5772 \ldots .$$

(γ is called Euler's constant), we have

$$\lim_{n \to \infty} \left(\Delta_n - \frac{2}{\pi} \log n \right) = \frac{2}{\pi} \left(\log \frac{8}{\pi} + \gamma \right) = 0.9625 \ldots . \tag{1.54}$$

THEOREM 1.2

For $n = 1, 2, \ldots, \Lambda_n(T) = \lambda_n(T; 1)$ and

$$\frac{2}{\pi} \log n + \frac{2}{\pi} \left(\log \frac{8}{\pi} + \gamma \right) < \Lambda_n(T) \leq \frac{2}{\pi} \log n + 1. \qquad (1.55)$$

Moreover, $\tau_n = \Lambda_n(T) - \dfrac{2}{\pi} \log n, n = 1, 2, \ldots,$ is a strictly monotone decreasing

sequence with $\tau_1 = 1$ and

$$\lim_{n \to \infty} \tau_n = \frac{2}{\pi} \left(\log \frac{8}{\pi} + \gamma \right).$$

Proof.   We show first that $\Lambda_n(T) = \lambda_n(T; 1) = \Delta_n$. If $n = 1$, this is trivial. Suppose $n \geq 2$. As we have seen (p. 16), $d_k(\theta) + d_{2n-k+1}(\theta)$ is an even function, hence a cosine polynomial. Thus, if $x = \cos \theta$, we have for $k = 1, \ldots, n,$

$$p_k(x) = d_k(\theta) + d_{2n-k+1}(\theta) \in \mathscr{P}_n.$$

Now for $i = 1, \ldots, n,$

$$p_k(\xi_i^{(n)}) = d_k(\theta_i) + d_{2n-k+1}(\theta_i) = d_k(\theta_i) = \begin{cases} 0, & i \neq k \\ 1, & i = k \end{cases},$$

and (1.46) reveals that the leading coefficient of $p_k(x)$ is zero. Hence

$$p_k(x) = l_{k,n}(T; x) \qquad k = 1, \ldots, n,$$

and

$$\lambda_n(T; x) = \sum_{k=1}^{n} |p_k(x)|.$$

Thus

$$\lambda_n(T; x) \leq \delta_n(\theta) \leq \Delta_n,$$

and it is easy to see from (1.43) that $\lambda_n(T; 1) = \Delta_n$. Hence $\Lambda_n(T) = \Delta_n$.

The rest of the theorem is proved by showing that the sequence $\tau_n = \Delta_n - (2/\pi) \log n$ is monotone decreasing as n increases. To this end we need some information about the monotone convergence of Riemann sums to the integral, and so we digress from the proof to obtain the following lemma due to D. J. Newman and the author, which is not without interest in itself.

LEMMA 1.2.1

If $f''(x)$ and $f'''(x)$ are both nonnegative in $[0, 1]$, the Riemann sums

$$b_n = \frac{1}{n} \sum_{k=1}^{n} f\left(\frac{2k-1}{2n}\right) \tag{1.56}$$

are monotone increasing as n increases.

Proof. Integrating three times by parts yields

$$\frac{1}{n} \sum_{k=1}^{n} f\left(\frac{2k-1}{2n}\right) = \int_0^1 f(t)\, dt - \frac{f''(0)}{24n^2}$$

$$- \int_0^1 \frac{4(nt - [nt + \tfrac{1}{2}])^3 + [nt + \tfrac{1}{2}]}{24n^3} f'''(1-t)\, dt, \tag{1.57}$$

(the [ ] here is the integer part notation; cf. footnote p. 2). Since $f''(0) \geq 0$, the sequence $-f''(0)/(24n^2)$ is monotone increasing; hence, since $t^3 f'''(1-t) \geq 0$, it suffices to show that the function

$$\frac{4(nt - [nt + \tfrac{1}{2}])^3 + [nt + \tfrac{1}{2}]}{24n^3 t^3}$$

decreases as n increases. Thus it is enough to show that

$$\frac{4(x - [x + \tfrac{1}{2}])^3 + [x + \tfrac{1}{2}]}{x^3}$$

is a decreasing function for $x > 0$. This function is continuously differentiable even at the points $k - (\tfrac{1}{2})$, k an integer; hence it suffices to verify that its derivative is negative for $k - (\tfrac{1}{2}) < x < k + (\tfrac{1}{2})$. In this interval, however, the function is

$$\frac{4(x - k)^3 + k}{x^3}$$

whose derivative is

$$\frac{12k}{x^4}((x - k)^2 - \tfrac{1}{4}),$$

which is indeed negative throughout the interval.

Returning now to the theorem, we apply the lemma with

$$f(x) = \frac{1}{(\pi/2)x} - \cot\frac{\pi}{2}x.$$

Since

$$\frac{1}{z} - \cot z = c_1 z + c_3 z^3 + \cdots + c_{2k-1} z^{2k-1} + \cdots,$$

with $c_{2k-1} > 0, k = 1, 2, \ldots$ (the expansion is valid in $|z| < \pi$. See Knopp [1]). It is evident that f'' and f''' are nonnegative in $[0, 1]$. The b_n defined in (1.56) satisfy $a_n = -(\pi/2)b_n$, where a_n is defined in (1.53). Since the b_n are monotone increasing, the a_n are monotone decreasing. Also, if we put

$$u_n = 2 \sum_{k=1}^{n} \frac{1}{2k-1} - \log n,$$

then

$$u_n - u_{n+1} = \log\left(1 + \frac{1}{n}\right) - \frac{2}{2n+1}$$

is positive for $n = 1$ and tends to zero as n tends to infinity.

Since the derivative of

$$\log\left(1 + \frac{1}{x}\right) - \frac{2}{2x+1}$$

is

$$-\frac{1}{x(x+1)(2x+1)^2} < 0, \qquad x > 0,$$

$u_n > u_{n+1}, n = 1, 2, \ldots$. Thus the u_n are strictly monotone decreasing, and so is the sequence $(\pi/2)\tau_n = a_n + u_n$. ∎

EXERCISES 1.3

1.3.1. For any X and $n = 1, 2, \ldots,$

$$\sum_{j=1}^{n} l_{j,n}(X; x) = 1, \tag{1.58}$$

and so

$$\lambda_n(X; x) \geqslant 1, \qquad n = 1, 2, \ldots, x \in I. \tag{1.59}$$

Hint. $1 \in P_n, n = 1, 2, \ldots$.

1.3.2. $$\frac{T_n(x)}{T_{n+1}(x)} = \frac{1}{n+1} \sum_{j+1}^{n+1} \frac{1 - (\xi_j^{(n+1)})^2}{x - \xi_j^{(n+1)}}.$$

1.3.3. Show that

$$L_{k-1}(T_n, T; x) = (-1)^r T_{|m|}(x),$$

where

$$n = 2kr + m; \quad -(k-1) \leqslant m \leqslant k, \quad r \geqslant 0.$$

Hint. $L_{k-1}(T_n, T; x) = (-1)^r L_{k-1}(T_{|m|}, T; x)$, according to Exercise 1.2.5. Now consider the two cases $-(k-1) \leqslant m \leqslant (k-1)$ and $m = k$.

1.3.4. If U denotes the array of nodes whose $n + 1$st row is $\eta_0, \ldots, \eta_n$, the extrema of $T_n(x)$, show that

$$L_n(f, U; x) = (1 - x^2) T_n'(x) \left\{ \frac{f(1)}{(1 - x) 2n^2} + \frac{f(-1)}{(1 + x)(-1)^{n+1} 2n^2} \right.$$

$$\left. + \sum_{j=1}^{n-1} \frac{f(\eta_j)}{(x - \eta_j)(1 - \eta_j^2) T_n''(\eta_j)} \right\}. \tag{1.60}$$

1.3.5. For any X and $n \geqslant 2$ show that

$$l_j'(x_j) = \frac{\omega''(x_j)}{2\omega'(x_j)}.$$

1.3.6. If $x_1 > x_2 > \cdots > x_n$, show that $\omega(x) = (x - x_1) \cdots (x - x_n)$ satisfies

$$\operatorname{sgn} \omega'(x_j) = (-1)^{j-1}, \quad j = 1, \ldots, n,$$

where

$$\operatorname{sgn} t = \begin{cases} 1, & t > 0, \\ -1, & t < 0, \\ 0, & t = 0. \end{cases}$$

If $x_1, x_2, \ldots, x_n$ are distinct real points and $f(x)$ is a function defined for $x = x_i, i = 1, \ldots, n$, the coefficient of x^{n-1} in the polynomial of degree at most $n - 1$, which interpolates f at the x_i, is denoted by $f(x_1, \ldots, x_n)$ and called the *divided difference of f with respect to $x_1, \ldots, x_n$.* Note that there is no notational ambiguity when $n = 1$.

1.3.7. Show that

$$f(x_1, \ldots, x_n) = \sum_{i=1}^{n} \frac{f(x_i)}{\omega'(x_i)}.$$

1.3.8. Show that if $x_1 \neq x_k$

$$\frac{f(x_1, \ldots, x_{k-1}) - f(x_2, \ldots, x_k)}{x_1 - x_k} = f(x_1, \ldots, x_k)$$

(hence the name divided difference).

1.3.9. Show that

$$p(x) = f(x_1) + (x - x_1)f(x_1, x_2) + (x - x_1)(x - x_2)f(x_1, x_2, x_3)$$
$$+ \cdots + (x - x_1) \cdots (x - x_{n-1})f(x_1, \ldots, x_n) \tag{1.61}$$

satisfies $p(x_i) = f(x_i)$, $i = 1, \ldots, n$. Equation 1.61 is called Newton's form of the interpolating polynomial.

Hint. Write the unique interpolating polynomial in the form $a_1 + a_2(x - x_1) + \cdots + a_k(x - x_1) \cdots (x - x_{k-1})$ and recall the definition of divided differences.

1.3.10. Show that

$$f(t) - L_{n-1}(f; t) = (t - x_1) \cdots (t - x_n)f(x_1, \ldots, x_n, t)$$

holds for all t. (The right-hand side is defined as zero when $t = x_i$, $i = 1, \ldots, n$.)
Hint. Use Exercise 1.3.7.

1.3.11. If $x_1, \ldots, x_n$ and t are points of $[a, b]$ and $f \in C^n[a, b]$, show that

$$f(t) - L_{n-1}(f; t) = (t - x_1) \cdots (t - x_n)\frac{f^{(n)}(\xi)}{n!}$$

for some $\xi(t)$ in $[a, b]$.

Hint. Let $h(t)/g(t) = (f(t) - L_n(f; t))/(t - x_1) \cdots (t - x_n)$. Then $h(t)g(x) - g(t)h(x)$, as a function of x has $n + 1$ zeros $x_1, \ldots, x_n, t$. Now apply Rolle's theorem.

1.3.12. If $x_1, \ldots, x_k$ are distinct points of $[a, b]$ and $f \in C^{k-1}[a, b]$, then

$$f(x_1, \ldots, x_k) = \frac{f^{(k-1)}(\xi)}{(k-1)!}$$

for some point ξ of $[a, b]$.

1.3.13. If $t \in \mathscr{T}_n$ (the trigonometric polynomials of degree at most n) and $t \neq 0$, show that t has at most $2n$ zeros in the interval $[0, 2\pi)$.

Hint. $t(\theta) = e^{-in\theta} q(e^{i\theta})$, where q is a polynomial of degree at most $2n$ with complex coefficients.

1.3.14. Is the t satisfying (1.49) unique?

1.3.15. Show that Lemma 1.2.1 is also valid if $f''(x) \geq 0$ and $f'''(x) \leq 0$.

1.3.16. Show that $\Lambda_n(T)$ is strictly monotone increasing with n.

1.3.17. If $t \in \mathscr{T}_n$, $0 \leq \alpha < \beta < \pi$, and $t(\alpha) t(\beta) > 0$ show that t has an even number (counting multiplicities) of zeros in (α, β). If $t(\alpha) t(\beta) < 0$ show that t has an odd number of zeros in (α, β).

1.3.18. If $t \in \mathcal{T}_n$, $0 \leqslant \phi_0 < \phi_1 < \cdots < \phi_k < 2\pi$, and $t(\phi_i)\, t(\phi_{i+1}) \leqslant 0$, $i = 0, \ldots,$ $k - 1$, show that t has at least k zeros in $[\phi_0, \phi_k]$.

Hint. Use Exercise 1.3.17.

1.3.19. Suppose that $t \in \mathcal{T}_n$ satisfies

$$\max_{0 \leqslant \theta < 2\pi} |t(\theta)| = 1$$

and $t(0) = 1$. Show that $t(\theta) \neq 0$ for $|\theta| < \pi/2n$, and $t(\pm\pi/(2n)) = 0$, if and only if $t(\theta) = \cos n\theta$.

Hint. Consider $r(\theta) = \cos n\theta - t(\theta)$ and suppose that the result we seek is false. Note that $r(j\pi/n)\, r((j+1)\pi/n) \leqslant 0$, $j = 1, \ldots, 2n - 2$ and $r(\pi/n) \leqslant 0$, whereas $r(\theta_0) > 0$, where θ_0 is a zero of t in $(0, \pi/(2n))$, say. Apply Exercise 1.3.18 and also observe that $r(0) = r'(0) = 0$.

1.3.20. (M. Riesz [1]) Suppose that $t \in \mathcal{T}_n$ and

$$\max_{0 \leqslant \theta < 2\pi} |t(\theta)| = M > 0.$$

If $|t(\phi)| = M$, then $t(\theta)$ has no zero in $|\theta - \phi| < \pi/(2n)$ and $t(\phi \pm \pi/(2n)) = 0$ if and only if $t(\theta) = \pm M \cos n(\theta - \phi)$.

1.3.21. Ehlich and Zeller [2] showed that for *even n*

$$\Lambda_n(U) = \Lambda_{n-1}(T).$$

Show that $\Lambda_{2k}(U) - (2/\pi) \log (2k)$ is a strictly monotone-*increasing* function of k with limit $(2/\pi)(\log(8/\pi) + \gamma)$.

Hint. Use (1.57) with $f(x) = |(\pi/2)x|^{-1} - \cot(\pi/2)x$ to obtain an upper bound on $b_{n+2} - b_n$, in the notation of (1.56).

1.4. Hermite Interpolation

We have just seen the sense in which the zeros of the Chebyshev polynomials are "good" nodes for polynomial interpolation. Our discussion was intended mainly as an introduction to the notion of polynomial interpolation, and the conclusion that there was no universal array of nodes at which the interpolating polynomials converged to every continuous function was disappointingly negative. To obtain positive results we must amplify the idea of interpolation. We do this by requiring the polynomial not only to take on given values at the nodes but by also fixing the value of its first derivative at the given nodes.

Given nodes $x_1, \ldots, x_n$, all contained in I, real numbers $y'_1, \ldots, y'_n$, and a function $f(x)$ defined on I, we wish to construct a $p \in \mathcal{P}_{2n-1}$ having the properties that

$$p(x_j) = f(x_j) = y_j, \qquad j = 1, \ldots, n \tag{1.62}$$

and

$$p'(x_j) = y_j', \qquad j = 1, \ldots, n. \tag{1.63}$$

Note that y_j' is not necessarily related to $f'(x_j)$, even if the latter exists. We put

$$\omega(x) = (x - x_1)(x - x_2) \cdots (x - x_n)$$

and construct two sets of fundamental polynomials,

$$h_j(x) = \left(1 - \frac{\omega''(x_j)}{\omega'(x_j)}(x - x_j)\right) l_j^2(x), \qquad j = 1, \ldots, n,$$

$$\mathfrak{h}_j(x) = (x - x_j) l_j^2(x), \qquad j = 1, \ldots, n, \tag{1.64}$$

where $l_j(x)$ is defined in (1.25). $h_1, \ldots, h_n \in \mathscr{P}_{2n-1}$ are called fundamental polynomials of the *first kind* for *Hermite interpolation* [which is the name given polynomial interpolation that satisfies (1.62) and (1.63)], and $\mathfrak{h}_1, \ldots, \mathfrak{h}_n \in \mathscr{P}_{2n-1}$ are fundamental polynomials of the *second kind* for Hermite interpolation. It is not hard to verify that

$$h_j(x_i) = \begin{cases} 0, & i \neq j \\ 1, & i = j \end{cases} \qquad i, j = 1, \ldots, n, \tag{1.65}$$

$$h_j'(x_i) = 0, \qquad i, j = 1, \ldots, n, \tag{1.66}$$
$$\mathfrak{h}_j(x_i) = 0, \qquad i, j = 1, \ldots, n, \tag{1.67}$$

$$\mathfrak{h}_j'(x_i) = \begin{cases} 0, & i \neq j \\ 1, & i = j \end{cases} \qquad i, j = 1, \ldots, n, \tag{1.68}$$

and the reader is urged to do so. Exercise 1.3.5 is useful in calculating $h_j'(x_j)$.
    The polynomial

$$W_{2n-1}(x) = \sum_{j=1}^{n} y_j h_j(x) + \sum_{j=1}^{n} y_j' \mathfrak{h}_j(x) \tag{1.69}$$

is a member of $\mathscr{P}_{2n-1}$ and satisfies

$$W_{2n-1}(x_j) = y_j,$$
$$W_{2n-1}'(x_j) = y_j',$$

in view of (1.65)–(1.68). The process of obtaining W_{2n-1} is called *Hermite interpolation*. Moreover, W_{2n-1} is the only member of $\mathscr{P}_{2n-1}$ with these properties, for if $p \in \mathscr{P}_{2n-1}$ satisfies (1.62) and (1.63) then $q = p - W_{2n-1} \in \mathscr{P}_{2n-1}, q(x_j) = 0, j = 1, \ldots, n$ and $q'(x_j) = 0, j = 1, \ldots, n$. Thus q has zeros of multiplicity at least 2 at $x_1, \ldots, x_n$, hence has, at least, $2n$ zeros and therefore $q = 0$. A consequence of the uniqueness of W_{2n-1} is that if $p \in \mathscr{P}_{2n-1}$ then

$$p(x) = \sum_{j=1}^{n} p(x_j) h_j(x) + \sum_{j=1}^{n} p'(x_j) \mathfrak{h}_j(x). \qquad (1.70)$$

Let us suppose now that $x_1, \ldots, x_n$ are chosen to be $\xi_1, \ldots, \xi_n$, the zeros of $T_n(x)$. Then, in view of Exercises 1.2.2 and 1.2.6, we obtain

$$h_j(x) = h_{j,n}(T; x) = \frac{1 - \xi_j x}{1 - \xi_j^2} (l_{j,n}(T; x))^2 = \frac{1 - \xi_j x}{n^2} \left(\frac{T_n(x)}{x - \xi_j} \right)^2 \qquad (1.71)$$

and

$$\mathfrak{h}_j(x) = \mathfrak{h}_{j,n}(T; x) = (x - \xi_j)(l_{j,n}(T; x))^2 = \frac{1}{n^2} \frac{1 - \xi_j^2}{x - \xi_j} T_n^2(x). \qquad (1.72)$$

EXERCISES 1.4

1.4.1. Show that for any choice of nodes

$$\sum_{j=1}^{n} h_j(x) = 1, \qquad (1.73)$$

and

$$\sum_{j=1}^{n} (x - x_j) h_j(x) = \sum_{j=1}^{n} \mathfrak{h}_j(x).$$

1.4.2. If we put

$$v_j(x) = 1 - \frac{\omega''(x_j)}{\omega'(x_j)} (x - x_j)$$

so that $h_j(x) = v_j(x) l_j^2(x)$, $j = 1, \ldots, n$, show that

$$\sum_{j=1}^{n} v_j(x) = n^2.$$

Hint. Recall Exercise 1.3.7.

1.4.3. Show that

$$h_{j,n}(T; x) \geqslant 0, \qquad x \in I, j = 1, \ldots, n. \qquad (1.74)$$

1.4.4. Show that

$$|\mathfrak{h}_{j,n}(T;x)| \leqslant h_j(x), \qquad x \in I, \qquad j = 1, \ldots, n, \tag{1.75}$$

hence that

$$\sum_{j=1}^{n} |\mathfrak{h}_{j,n}(T;x)| \leqslant 1, \tag{1.76}$$

1.4.5. Show that

$$\sum_{j=1}^{n} \mathfrak{h}_{j,n}(T;x) = \frac{1}{n} T_{n-1}(x) \, T_n(x).$$

1.4.6. If $p \in \mathscr{P}_{2n-1}$ and

$$|p(\xi_j)| \leqslant A, \qquad |p'(\xi_j)| \leqslant B; \qquad j = 1, \ldots, n,$$

then

$$|p(x)| \leqslant A + B, \qquad x \in I. \tag{1.77}$$

If $n = 1$, the bound $A + B$ in (1.77) cannot be improved.

Hint. Apply (1.70), (1.73), (1.74), and (1.76).

1.4.7. Equation 1.38 shows that (1.76) can be improved to

$$\sum_{j=1}^{n} \frac{|\mathfrak{h}_{j,n}(T;x)|}{(1 - \xi_j^2)^{1/2}} \leqslant \frac{1}{n}\left(\frac{2}{\pi} \log n + 1\right). \tag{1.78}$$

1.4.8. Prove that under the hypotheses of Exercise 1.4.6 the conclusion (1.77) can be strengthened to read

$$|p(x)| \leqslant A + \mu_n B, \qquad x \in I, \tag{1.79}$$

where

$$\mu_n = \frac{1}{n}\left(\frac{2}{\pi} \log n + 1\right).$$

Indeed, show that (1.79) remains true if the hypothesis on p' in Exercise 1.4.6 is weakened to read

$$|p'(\xi_j)| \leqslant B(1 - \xi_j^2)^{-1/2}, \qquad j = 1, \ldots, n.$$

1.4.9. Show that

$$\sum_{j=1}^{n} [l_{j,n}(T;x)]^2 \leqslant 2 \cos^2 \frac{\pi}{4n} \leqslant 2, \qquad \text{all} \qquad x \in I.$$

Hint. Use (1.73) and (1.71). (Compare this result with (1.37), taking $X = T$.)

1.4.10. Show that for $j = 1, \ldots, n$

$$\max_{-1 \leqslant x \leqslant 1} |l_{j,n}(T; x)| \leqslant \sqrt{2}.$$

We show next that Hermite interpolation in the Chebyshev nodes succeeds where Lagrange interpolation failed; i.e., given a continuous function it provides us with a sequence of polynomials that converges to the function. This result is due to L. Fejér [1].

THEOREM 1.3

Let $f(x)$ be continuous on I. Let $W_{2n-1}(x)$ be the Hermite interpolating polynomial defined by the conditions

$$W_{2n-1}(\xi_j) = y_j = f(\xi_j), \qquad j = 1, \ldots, n, \tag{1.80}$$

$$W'_{2n-1}(\xi_j) = 0, \qquad j = 1, \ldots, n; \tag{1.81}$$

then

$$\lim_{n \to \infty} W_{2n-1}(x) = f(x) \tag{1.82}$$

uniformly in I.

Proof. In view of $(1.69)(y'_j = 0)$, we have

$$W_{2n-1}(x) = \sum_{j-1}^{n} f(\xi_j)\, h_{j,n}(T; x)$$

and, recalling (1.73),

$$f(x) = \sum_{j=1}^{n} f(x)\, h_{j,n}(T; x).$$

Hence for $x \in I$

$$|f(x) - W_{2n-1}(x)| \leqslant \sum_{j=1}^{n} |f(x) - f(\xi_j)|\, h_{j,n}(T; x), \tag{1.83}$$

since $h_{j,n}(T; x) \geqslant 0, j = 1, \ldots, n$ (cf. Exercise 1.4.3).

Given $\epsilon > 0$, let us choose $\delta > 0$ and so small that

$$|f(x') - f(x'')| < \frac{\epsilon}{2}$$

whenever $|x' - x''| < \delta$, $x', x'' \in I$. This can be done, for f is uniformly continuous on I. Fix $x \in I$ and let α be the set of j for which $|\xi_j - x| < \delta$; β denotes the set of the remaining j among $1, \ldots, n$. Then, in view of (1.74) and (1.73),

$$\sum_{j \in \alpha} |f(x) - f(\xi_j)| \, h_{j,n}(T;x) < \frac{\epsilon}{2} \sum_{j \in \alpha} h_{j,n}(T;x)$$

$$\leqslant \frac{\epsilon}{2} \sum_{j=1}^{n} h_{j,n}(T;x) = \frac{\epsilon}{2}. \tag{1.84}$$

Moreover, if $j \in \beta$, then $|\xi_j - x| \geqslant \delta$, and so

$$h_{j,n}(T;x) = \frac{1 - \xi_j x}{n^2} \left(\frac{T_n(x)}{x - \xi_j} \right)^2 < \frac{2}{n^2 \delta^2},$$

since $1 - \xi_j x < 2$ and $T_n^2(x) \leqslant 1$. If

$$M = \max_{-1 \leqslant x \leqslant 1} |f(x)|,$$

then $|f(x) - f(\xi_j)| \leqslant 2M$, $j = 1, \ldots, n$, and

$$\sum_{j \in \beta} |f(x) - f(\xi_j)| \, h_{j,n}(T;x) < \frac{4M}{n\delta^2}, \tag{1.85}$$

since the number of indices in β does not exceed n.

From (1.83) and (1.85) we conclude that for each $x \in I$

$$|f(x) - W_{2n-1}(x)| < \frac{\epsilon}{2} + \frac{4M}{n\delta^2},$$

and so there exists N such that for $n > N$

$$|f(x) - W_{2n-1}(x)| < \epsilon;$$

i.e.,

$$\lim_{n \to \infty} W_{2n-1}(x) = f(x)$$

uniformly in I. ■

An immediate consequence of Theorem 1.3 is the Weierstrass approximation theorem.

THEOREM 1.4

Given $f(x)$ continuous on I and $\epsilon > 0$, there exists a polynomial, $p(x)$, such that

$$|f(x) - p(x)| < \epsilon$$

for all $x \in I$.

The Weierstrass theorem is the theoretical basis for the great utility of polynomials, since, roughly speaking, it enables us to replace any continuous function with a polynomial in the course of a mathematical argument.

Theorem 1.3 was proved by Fejér in 1916. In 1930 (Fejér [2]) he returned to the same topic and was able to improve the result by weakening the requirement in (1.81) that the derivative of the interpolating polynomial vanish at the Chebyshev nodes. More precisely he proved the following:

THEOREM 1.5

Let $f(x)$ be continuous on I. Let $W_{2n-1}(x)$ be a Hermite interpolating polynomial defined by the conditions

$$W_{2n-1}(\xi_j) = f(\xi_j), \qquad j = 1, \ldots, n,$$
$$W'_{2n-1}(\xi_j) = y'_j, \qquad j = 1, \ldots, n,$$

where

$$|y'_j| \leqslant \epsilon_n \frac{n}{\log n} (1 - \xi_j^2)^{-1/2}, \qquad j = 1, \ldots, n, \tag{1.86}$$

with

$$\lim_{n \to \infty} \epsilon_n = 0. \tag{1.87}$$

Then $\lim_{n \to \infty} W_{2n-1}(x) = f(x)$ uniformly in I.

Proof

$$W_{2n-1}(x) = \sum_{j=1}^{n} f(\xi_j) h_j(x) + \sum_{j=1}^{n} y'_j \mathfrak{h}_j(x),$$

but by Theorem 1.3 we know that

$$\lim_{n \to \infty} \left(\sum_{j=1}^{n} f(\xi_j) \, h_j(x) \right) = f(x)$$

uniformly in I, whereas in view of (1.78) and (1.86)

$$\left| \sum_{j=1}^{n} y'_j \, \mathfrak{h}_j(x) \right| \leqslant \epsilon_n \left(\frac{2}{\pi} + \frac{1}{\log n} \right)$$

and

$$\epsilon_n\left(\frac{2}{\pi} + \frac{1}{\log n}\right) \to 0$$

as $n \to \infty$, by (1.87), thus proving the theorem. ∎

EXERCISES

1.4.11. Prove that if $f'(x)$ is bounded on I and W_{2n-1} satisfies

$$W_{2n-1}(\xi_j) = f(\xi_j), \qquad j = 1, \ldots, n,$$
$$W'_{2n-1}(\xi_j) = f'(\xi_j), \qquad j = 1, \ldots, n,$$

then $\lim_{n \to \infty} W_{2n-1}(x) = f(x)$ uniformly in I.

1.4.12. Prove that if $f'(x)$ is continuous on I the Weierstrass approximation theorem (Theorem 1.4) can be strengthened by adding the conclusion that $|f'(x) - p'(x)| < \epsilon$ for all $x \in I$.

Hint. Apply Theorem 1.4 to $f'(x)$ and consider the indefinite integral of the polynomial thus obtained.

1.5. Orthogonality

Further interesting properties of the Chebyshev polynomials follow directly from the definition (1.2). It is easy to verify that for all nonnegative integers m, k

$$\int_0^\pi \cos k\theta \, \cos m\theta \, d\theta = 0, \qquad m \neq k, \tag{1.88a}$$

$$\int_0^\pi \cos^2 k\theta \, d\theta = \begin{cases} \dfrac{\pi}{2}, & k \neq 0, \\ \pi, & k = 0. \end{cases} \tag{1.88b}$$

If we make the change of variables $x = \cos\theta$ in (1.88a, b), we obtain the *orthogonality relationship*

$$\int_{-1}^1 T_k(x) \, T_m(x) \, \frac{dx}{\sqrt{1-x^2}} = 0, \qquad m \neq k, \tag{1.89a}$$

$$\int_{-1}^1 T_k^2(x) \, \frac{dx}{\sqrt{1-x^2}} = \begin{cases} \dfrac{\pi}{2}, & k \neq 0, \\ \pi, & k = 0; \end{cases} \tag{1.89b}$$

that is to say, the Chebyshev polynomials $\{T_n(x)\}_{n=0}^\infty$ form a sequence of *orthogonal polynomials* on I with respect to the *weight function* $(1-x^2)^{-1/2}$.

As such they are members of several large, important, and much studied families of sequences of orthogonal polynomials:

1. Sequences of polynomials $\{p_n(x)\}_{n=0}^{\infty}$ that satisfy

$$\int_{-1}^{1} p_k(x)\, p_m(x)\, w(x)\, dx = 0, \qquad m \neq k, \tag{1.90}$$

with a weight function $w(x) \geqslant 0$ on I.

2. The subset of (1) consisting of sequences of polynomials $\{p_n^{(\alpha,\beta)}(x)\}_{n=0}^{\infty}$ that satisfy (1.90) with

$$w(x) = (1 - x)^{\alpha}(1 + x)^{\beta}, \qquad \alpha > -1, \beta > -1. \tag{1.91}$$

These are called the *Jacobi* polynomials.

3. The subset of (2) consisting of sequences of polynomials $\{p_n^{(\lambda)}(x)\}_{n=0}^{\infty}$ that satisfy (1.90) and (1.91) with $\alpha = \beta$ and $\lambda = \alpha + \frac{1}{2}$. These are called the *ultraspherical* (or *Gegenbauer*) polynomials.

It is clear that the Chebyshev polynomials are ultraspherical polynomials with $\lambda = 0$. We shall examine some properties of the Chebyshev polynomials that are characteristic of the larger classes of orthogonal polynomials mentioned above. The reader who is interested in seeing the generalizations of these results to the larger classes (and learning to which class a specific result generalizes) of orthogonal polynomials should consult Szegö [1].

Second-Order Linear Homogeneous Differential Equation

We saw in (1.23) that

$$\frac{1}{n} T_n'(x) = \frac{\sin n\theta}{\sin \theta}, \qquad x = \cos \theta.$$

Therefore

$$T_n''(x) = n \frac{d}{d\theta}\left(\frac{\sin n\theta}{\sin \theta}\right)\left(-\frac{1}{\sin \theta}\right),$$

from which it is easy to verify that $y = T_n(x)$ satisfies the second-order linear homogeneous differential equation

$$(1 - x^2)y'' - xy' + n^2 y = 0 \tag{1.92}$$

for $x \in I$, hence for all x.

If we write [cf. (1.9)]

$$T_n(x) = t_0 + t_1 x + \cdots + t_n x^n$$

and substitute in (1.92), we obtain

$$(1 - x^2) \sum_{k=0}^{n} k(k-1)t_k x^{k-2} - x \sum_{k=0}^{n} k t_k x^{k-1} + n^2 \sum_{k=0}^{n} t_k x^k = 0,$$

or

$$\sum_{k=0}^{n} k(k-1)t_k x^{k-2} + \sum_{k=0}^{n} (n^2 - k^2) t_k x^k = 0.$$

Combining coefficients of like powers gives

$$0 = \sum_{k=0}^{n-2} (t_k(n^2 - k^2) + t_{k+2}(k+2)(k+1))x^k + (n^2 - (n-1)^2)t_{n-1}x^{n-1},$$

from which we conclude that

$$t_{n-1} = 0 \tag{1.93}$$

and

$$t_k(n^2 - k^2) + t_{k+2}(k+2)(k+1) = 0, \qquad k = 0, \ldots, (n-2). \tag{1.94}$$

Equations and 1.94 immediately imply [what we already knew; cf. (1.10)] that $t_{n-(2k-1)} = 0$. Since, according to (1.14),

$$t_n = 2^{n-1}$$

we see that

$$t_{n-2} = -2^{n-3} \cdot \frac{n(n-1)}{(n-1)}$$

and

$$t_{n-4} = \frac{n(n-1)(n-2)(n-3)}{2!(n-1)(n-2)} 2^{n-5}.$$

In general we have

$$t_{n-2m} = (-1)^m \frac{n(n-1) \cdots (n-2m+1)}{m!(n-1)(n-2) \cdots (n-m)} 2^{n-2m-1}, \tag{1.95}$$

as we may readily establish by mathematical induction. From (1.95) we easily obtain

$$t_{n-2m}^{(n)} = (-1)^m \frac{n}{n-m} \binom{n-m}{m} 2^{n-2m-1}, \qquad m = 0, 1, \ldots, \left[\frac{n}{2}\right], \tag{1.96}$$

a more concise form for the nonzero coefficients than we had before.

EXERCISE 1.5

1.5.1. Show that if $-1 \leqslant x \leqslant 1$ and $n \neq 0$

$$\int_{-1}^{x} T_n(t) \frac{dt}{\sqrt{1-t^2}} = -\frac{\sqrt{1-x^2}\, U_{n-1}(x)}{n},$$

hence

$$\left| \int_{-1}^{x} T_n(t) \frac{dt}{\sqrt{1-t^2}} \right| \leqslant \frac{1}{n}.$$

1.5.2. Show that for $n \geqslant 1$

$$\int_{-1}^{1} |T_n(t)| \frac{dt}{\sqrt{1-t^2}} = 2.$$

1.5.3. Show that for $n \geqslant 2$

(a) $I_n = \int_{-1}^{1} |T_n(x)|\, dx = \frac{2}{n^2-1}\left[\frac{n}{\sin \pi/2n} - 1 \right].$

(b) $I_1, I_2, I_3, \ldots,$ is a monotone-increasing sequence with limit $4/\pi$.

1.5.4. If the polynomial $p\ (\neq 0)$ satisfies the differential equation (1.92), then $p = cT_n$ for some constant c.

Hint. Putting $y = p$ in (1.92) yields a polynomial identity. Examine the leading coefficient on the left and thereby determine the degree of p.

1.5.5. Show that for $k \geqslant 1$

$$(1-x^2)T_n^{(k+1)}(x) - (2k-1)\,xT_n^{(k)}(x) + (n^2 - (k-1)^2)T_n^{(k-1)}(x) = 0.$$

1.5.6. Verify that

$$T_n^{(k)}(1) = \frac{n^2(n^2-1)(n^2-2^2)\cdots(n^2-(k-1)^2)}{1.3.5\cdots(2k-1)}. \tag{1.97}$$

1.5.7. Show that (1.60) may be simplified and written

$$L_n(f, U; x) = \frac{(1-x^2)T_n'(x)}{n^2} \sum_{j=0}^{n}{}''(-1)^{j+1}\frac{f(\eta_j)}{(x-\eta_j)}, \tag{1.98}$$

where Σ'' is a "trapezoidal" sum; i.e.,

$$\sum_{j=0}^{n}{}'' u_j = \tfrac{1}{2}u_0 + u_1 + u_2 + \cdots + u_{n-1} + \tfrac{1}{2}u_n. \tag{1.99}$$

1.5.8. (Duffin and Schaefer [1])   Prove that if $p \in \mathscr{P}_n$ and

$$|p(\eta_j^{(n)})| \leqslant 1, \qquad j = 0, \ldots, n,$$

then

$$|p'(\xi_j^{(n)})| \leqslant n(1 - (\xi_j^{(n)})^2)^{-1/2} = |T_n'(\xi_j^{(n)})|, \qquad j = 1, \ldots, n. \tag{1.100}$$

Equality in (1.100) occurs for any one j only if $p(x) = \pm T_n(x)$.

Hint.   Note that $p'(\xi_j) = L_n'(p, U; \xi_j)$; hence $T_n'(\xi_j) = L_n'(T_n, U; \xi_j)$ and recall Exercise 1.2.3.

1.5.9. (Duffin and Schaeffer [1])   Prove that if $p \in \mathscr{P}_n$ and

$$|p(\eta_j^{(n)})| \leqslant 1, \qquad j = 0, \ldots, n,$$

then

$$|p^{(k)}(\tau)| \leqslant |T_n^{(k)}(\tau)|, \qquad k = 1, \ldots, n,$$

where τ is any zero of $T_n^{(k-1)}(x)$. Equality occurs for any single τ only if $p = \pm T_n(x)$.

Hint.   Note that $k = 1$ is Exercise 1.5.8. Use mathematical induction on k and express $p^{(k)}$ as its own interpolating polynomial in the zeros of $T_n^{(k-1)}$.

1.5.10.   If $p \in \mathscr{P}_n$ and

$$|p(\eta_j^{(n)})| \leqslant 1, \qquad j = 0, \ldots, n,$$

show that

$$|p^{(k)}(x)| \leqslant T_n^{(k)}(x), \qquad x \geqslant u,$$

where u is the largest zero of $T_n^{(k-1)}(x)$.

1.5.11. (Rogosinski [1])   If $p \in \mathscr{P}_n$ and

$$|p(\eta_j^{(n)})| \leqslant 1, \qquad j = 0, \ldots, n,$$

then

$$|p^{(k)}(t)| \leqslant |T_n^{(k)}(t)|, \quad |t| \geqslant 1, \qquad k = 0, \ldots, n,$$

with equality only if $p = \pm T_n$ for $k \geqslant 1$ and $k = 0, |t| > 1$.

1.5.12.   (L. Harris)   If $x, y \geqslant 1$, then

$$T_n(xy) \leqslant T_n(x) T_n(y);$$

indeed, more generally, if $0 \leqslant j, k$ and $j + k \leqslant n$,

$$x^k T_n^{(j+k)}(xy) \leqslant T_n^{(j)}(x) T_{n-j}^{(k)}(y).$$

Hint. Apply Exercise 1.5.11 to $p(y) = T_n^{(j)}(xy)/T_n^{(j)}(x), x \geqslant 1$.

For generalizations see Askey, Gasper, and Harris [1].

1.5.13. Show that

$$U_n\left(\frac{x}{2}\right) = \sum_{k=0}^{[n/2]} (-1)^k \binom{n-k}{k} x^{n-2k}.$$

[$U_n(x)$ is defined in (1.23).]

Three-Term Recurrence Formula

The three-term recurrence formula satisfied by the Chebyshev polynomials is the translation of the elementary trigonometric identity

$$\cos n\theta + \cos(n-2)\theta = 2\cos\theta \cos(n-1)\theta,$$

which becomes

$$T_n(x) = 2xT_{n-1}(x) - T_{n-2}(x), \qquad n = 2, 3, \ldots, \tag{1.101}$$

with $T_0(x) = 1$ and $T_1(x) = x$.

EXERCISES 1.5 (continued)

1.5.14. Show that (1.101) is valid for $n = 0, \pm 1, \pm 2, \ldots$, if we put

$$T_{-n} = T_n$$

for positive integers n.

1.5.15. Show that no two consecutive Chebyshev polynomials, $T_k(x), T_{k+1}(x)$, have a zero in common.

1.5.16.

$$(1 - x^2) T_n'(x) = n[T_{n-1}(x) - xT_n(x)].$$

1.5.17. Show that

$$\sum_{j=0}^{n}{}' T_j(x) T_j(y) = \frac{1}{2}\left[\frac{T_{n+1}(x) T_n(y) - T_n(x) T_{n+1}(y)}{x - y}\right] \tag{1.102}$$

where $\sum_{j=0}^{\prime n} u_j$ means

$$\tfrac{1}{2}u_0 + u_1 + u_2 + \cdots + u_n.$$

1.5.18.   Show that

$$2 \sum_{j=1}^{n}{}' T_j^2 = T'_{n+1} T_n - T_{n+1} T'_n, \qquad (1.103)$$

hence that

$$T'_{n+1} T_n - T_{n+1} T'_n \geqslant 1. \qquad (1.104)$$

Also show that $T_{n+1}^2(x) - T_n(x) T_{n+2}(x)$ is positive, zero, or negative for $|x| < 1, |x| = 1$, and $|x| > 1$, respectively.

1.5.19.   Show that the Chebyshev polynomials of the second kind satisfy the three-term recurrence formula $U_n(x) = 2xU_{n-1}(x) - U_{n-2}(x), n = 2, 3, \ldots,$ [identical to (1.101)] with $U_0(x) = 1$ and $U_1(x) = 2x$.

Generating Function

Suppose that $|u| < 1$, then

$$\sum_{n=0}^{\infty} u^n e^{in\theta} = \sum_{n=0}^{\infty} (ue^{i\theta})^n = \frac{1}{1 - ue^{i\theta}}.$$

On equating the real parts of this equality, we obtain

$$\sum_{n=0}^{\infty} u^n \cos n\theta = \frac{1 - u \cos \theta}{1 + u^2 - 2u \cos \theta}$$

or

$$F(u, x) = \frac{1 - ux}{1 + u^2 - 2ux} = \sum_{n=0}^{\infty} T_n(x) u^n, \qquad x \in I. \qquad (1.105)$$

The function $F(u, x)$ is called a generating function for the Chebyshev polynomials, since they appear as the coefficients in its expansion in powers of u.

It is interesting to remark that we can recover (1.96) from (1.105). To do so we note that

$$F\left(u, \frac{x}{2}\right) = \left(1 - \frac{ux}{2}\right) \frac{1}{1 - u(x - u)}. \qquad (1.106)$$

If we suppose that $|u| \leqslant \frac{1}{2}$ and that $x \in I$, then

$$-\tfrac{3}{4} \leqslant u(x - u) \leqslant \tfrac{1}{4}$$

and so

$$\frac{1}{1 - u(x - u)} = \sum_{k=0}^{\infty} (u(x - u))^k = \sum_{k=0}^{\infty} u^k (x - u)^k. \qquad (1.107)$$

The coefficient of u^m in the right-most expression in (1.107) is

$$x^m - \binom{m-1}{1} x^{m-2} + \binom{m-2}{2} x^{m-4} + \cdots + (-1)^j \binom{m-j}{j} x^{m-2j}$$

$$+ \cdots + (-1)^{[m/2]} \binom{n - [m/2]}{[m/2]} x^{m-2[m/2]},$$

as we can see by starting with the term $k = m$ in the infinite series, extracting from it the term in u^m, then considering the term $k = m - 1$ in the infinite series, extracting from it the term in u^m, and so on. (As an aside, compare Exercise 1.5.13. We have stumbled on the generating function of $\{U_n(x/2)\}$.) Therefore the coefficient of u^n in the expansion $F(u, x/2)$ is, in view of (1.106) and (1.107),

$$T_n\left(\frac{x}{2}\right) = \sum_{k=0}^{[(n-1)/2]} (-1)^k \left[\binom{n-k}{k} - \frac{1}{2} \binom{n-1-k}{k} \right] x^{n-2k} + \cos\frac{n\pi}{2},$$

$$(1.108)$$

and since this equality holds for $x \in I$ it holds for all x; but

$$\binom{n-k}{k} - \frac{1}{2} \binom{n-1-k}{k} = \frac{1}{2} \frac{n-k}{k} \binom{n-k}{k},$$

and so replacing x with $2x$ in (1.108) enables us to recover (1.96).

Least Squares

We show next that if

$$p(x) = a_0 + a_1 x + \cdots + a_{n-1} x^{n-1} + x^n, \qquad n > 0$$

then

$$\int_{-1}^{1} p^2(x) \frac{dx}{\sqrt{1 - x^2}} \geqslant \int_{-1}^{1} \tilde{T}_n^2(x) \frac{dx}{\sqrt{1 - x^2}} = 2^{1-2n} \pi \qquad (1.109)$$

with equality only if $p = \tilde{T}_n$; that is to say, among all polynomials of degree n having leading coefficient 1 the normalized Chebyshev polynomial has the least integral of its square with respect to the weight function $(1 - x^2)^{-1/2}$.

To prove (1.109) we write $p(x) = b_0 T_0(x) + \cdots + b_n T_n(x)$, where (cf. Exercise 1.2.6) $b_n = 2^{-(n-1)}$, and consider

$$\int_{-1}^{1} (\tilde{T}_n(x) - p(x))^2 \frac{dx}{\sqrt{1 - x^2}} = \int_{-1}^{1} \tilde{T}_n^2(x) \frac{dx}{\sqrt{1 - x^2}} + \int_{-1}^{1} p^2(x) \frac{dx}{\sqrt{1 - x^2}}$$

$$- 2 \int_{-1}^{1} (b_0 + b_1 T_1(x) + \cdots + b_n T_n(x)) \tilde{T}_n(x) \frac{dx}{\sqrt{1 - x^2}}. \qquad (1.110)$$

But the orthogonality relationship (1.89) implies that the last term on the right-hand side of (1.110) is equal to

$$- 2 \int_{-1}^{1} \tilde{T}_n^2(x) \frac{dx}{\sqrt{1 - x^2}} .$$

Moreover, the left-hand side of (1.110) is nonnegative (since it is the integral of a nonnegative integrand) and zero if and only if $p = \tilde{T}_n$. Equation 1.109 follows at once, the value of the minimal integral following from (1.89b).

Numerical Integration

Numerical integration is approximation of the definite integral by finite sums. A typical numerical integration formula involves approximating

$$\int_{-1}^{1} f(x) \, dx \qquad (1.111)$$

by

$$\sum_{i=1}^{n} A_i^{(n)} f(x_i^{(n)}) \qquad (1.112)$$

for all f continuous on I. One criterion of the goodness of the approximation (1.112) to (1.111) is to require that (1.112) be equal to (1.111) for $f \in \mathscr{P}_k$ and k as large as possible. A reason for adopting this criterion is that, according to the Weierstrass approximation theorem (Theorem 1.4), every continuous function on I can be uniformly approximated, arbitrarily closely, by polynomials.

In order to integrate every $f \in \mathscr{P}_k$ exactly by means of (1.112) it suffices to choose the $A_i^{(n)}$ and $x_i^{(n)}$ in (1.112) to satisfy

$$\sum_{i=1}^{n} A_i^{(n)} (x_i^{(n)})^j = \int_{-1}^{1} x^j \, dx, \qquad j = 0, 1, \ldots, k. \tag{1.113}$$

We have $2n$ unknowns in the system of $k + 1$ equations (1.113). Therefore the largest k for which we can generally expect to solve this system is $k = 2n - 1$; hence we can expect to integrate exactly all $f \in \mathscr{P}_{2n-1}$, at most, by suitable choice of nodes and coefficients in (1.112).

The choice of the zeros of the Chebyshev polynomials as nodes in (1.112) leads to an optimal numerical integration formula, not, however, for (1.111) but for

$$\int_{-1}^{1} \frac{f(x)}{\sqrt{1 - x^2}} \, dx. \tag{1.114}$$

We proceed next to produce this formula. Suppose that $p \in \mathscr{P}_{2n-1}$; then the Lagrange interpolating polynomial to p at the zeros of $T_n(x)$ is

$$L_{n-1}(p, T; x) = \sum_{j=1}^{n} \frac{p(\xi_j) T_n(x)}{T_n'(\xi_j)(x - \xi_j)}. \tag{1.115}$$

Since $p(x) - L_{n-1}(x) = 0$ for $x = \xi_1, \ldots, \xi_n$, we can write

$$p(x) - L_{n-1}(x) = T_n(x) r(x), \tag{1.116}$$

where $r(x) \in \mathscr{P}_{n-1}$. In view of Exercise 1.2.6 and the orthogonality relationships (1.89a)

$$\int_{-1}^{1} T_n(x) r(x) \frac{dx}{\sqrt{1 - x^2}} - 0,$$

and so (1.116) implies

$$\int_{-1}^{1} p(x) \frac{dx}{\sqrt{1 - x^2}} = \int_{-1}^{1} L_{n-1}(p, T; x) \frac{dx}{\sqrt{1 - x^2}}, \tag{1.117}$$

but, if we evaluate the integral on the right-hand side of (1.117) in view of (1.115), we obtain

$$\int_{-1}^{1} \frac{p(x)}{\sqrt{1 - x^2}} \, dx = \sum_{j=1}^{n} \lambda_j^{(n)} p(\xi_j^{(n)}), \tag{1.118}$$

where

$$\lambda_j^{(n)} = \frac{1}{T_n'(\xi_j^{(n)})} \int_{-1}^{1} \frac{T_n(x)}{(x - \xi_j^{(n)})} \frac{dx}{\sqrt{1 - x^2}}.$$ (1.119)

In other words, the numerical integration formula

$$\sum_{j=1}^{n} \lambda_j^{(n)} f(\xi_j^{(n)}),$$ (1.120)

where $\lambda_j^{(n)}$ is defined in (1.119), evaluates the integral (1.114) exactly if $f \in \mathscr{P}_{2n-1}$. Formula 1.120 is simply the integral with respect to the weight function $(1 - x^2)^{-1/2}$ of $L_{n-1}(f, T; x)$.

The formula analogous to (1.120) for (1.111) is due to Gauss and called Gauss's *quadrature formula* (quadrature being a synonym of numerical integration). It is derived by using the zeros of the set of polynomials orthogonal on I with weight function $w(x) = 1$, the Legendre polynomials, in place of the zeros of the Chebyshev polynomials in (1.115). The reader is once again referred to Szegö [1] for details. Formula is 1.120 is sometimes called the Gauss-Chebyshev quadrature formula.

Formula 1.120 cannot evaluate (1.114) exactly for all $f \in \mathscr{P}_{2n}$, for if $f(x) = T_n^2(x) \in \mathscr{P}_n$ then (1.114) is positive and (1.120) is zero. Indeed, there is no formula

$$\sum_{i=1}^{n} c_i f(x_i), \qquad c_i \neq 0, \ i = 1, \ldots, n,$$ (1.121)

that evaluates (1.114) exactly for $f \in \mathscr{P}_m$ with $m \geq 2n - 1$ other than (1.120), for if there were put $\omega(x) = (x - x_1) \cdots (x - x_n)$. Say x_k is not one of the $\xi_j^{(n)}$. Consider

$$f(x) = \frac{\omega(x) T_n(x)}{x - x_k};$$

then $f \in \mathscr{P}_{2n-1}$, (1.120) is zero, and therefore (1.114) is also, but (1.121) yields the value $c_k \omega'(x_k) T_n(x_k) \neq 0$. Thus, possibly after renumbering, $x_j = \xi_j^{(n)}, j = 1, \ldots, n$. Finally, putting $f(x) = T_n(x)/(x - \xi_j)$ in (1.121) and (1.120) yields

$$c_j = \lambda_j^{(n)} = \frac{1}{T_n'(\xi_j)} \int_{-1}^{1} \frac{T_n(x)}{x - \xi_j} \frac{dx}{\sqrt{1 - x^2}}.$$

We want to show next that (1.119) can be considerably simplified. Indeed

$$\lambda_j^{(n)} = \frac{\pi}{n}, \qquad j = 1, \ldots, n, \tag{1.122}$$

so that (1.120) has the particularly simple form

$$\frac{\pi}{n} \sum_{j=1}^{n} f(\xi_j^{(n)}). \tag{1.123}$$

To prove (1.122) we put $y = \xi_i$ in (1.102), which then becomes

$$\sum_{j=0}^{n} T_j(x) T_j(\xi_i) = -\tfrac{1}{2} T_{n+1}(\xi_i) \frac{T_n(x)}{x - \xi_i}.$$

If we now multiply both sides of this identity by $(1 - x^2)^{-1/2}$ and then integrate over I, we obtain, in view of the orthogonality relationships,

$$\frac{\pi}{2} = -\tfrac{1}{2} T_{n+1}(\xi_l) \, T_n'(\xi_l) \lambda_l^{(n)}.$$

Equations 1.122 now follow from Exercises 1.2.3 and 1.2.7.

We observe, finally, that the approximation (1.123) to (1.114) converges to (1.114) as $n \to \infty$, for

$$\frac{\pi}{n} \sum_{j=1}^{n} f(\xi_j^{(n)}) = \frac{\pi}{n} \sum_{j=1}^{n} f\!\left(\cos{(2j - 1)}\frac{\pi}{2n}\right),$$

and since $f(x)$ is continuous on I the right-hand side of this equality converges to

$$\int_0^{\pi} f(\cos\theta) \, d\theta = \int_{-1}^{1} \frac{f(x)}{\sqrt{1 - x^2}} dx$$

as $n \to \infty$.

EXERCISES 1.5 (continued)

1.5.20. Equations 1.122 are equivalent to

$$\frac{\pi}{n} = \int_{-1}^{1} l_{j,n}(T; x) \frac{dx}{\sqrt{1 - x^2}} \tag{1.124}$$

(the $l_{j,n}(T;x)$ are the fundamental polynomials of Lagrange interpolation at the zeros of the Chebyshev polynomials). Show that

$$\int_{-1}^{1} l_{j,n}(T;x) l_{k,n}(T;x) \frac{dx}{\sqrt{1-x^2}} = 0, \qquad j \neq k; j, \quad k = 1, \ldots, n.$$

1.5.21.   Show that

$$\frac{\pi}{n} = \int_{-1}^{1} [l_{j,n}(T;x)]^2 \frac{dx}{\sqrt{1-x^2}}, \qquad j = 1, \ldots, n. \qquad (1.125)$$

Hint.   Substitute $p(x) = [l_{j,n}(T;x)]^2 \in \mathcal{P}_{2n-1}$ in (1.118). Note the remarkable result implied by (1.124) and (1.125). Also summing both sides of (1.125) on j from 1 to n gives

$$\int_{-1}^{1} \sum_{j=1}^{n} [l_{j,n}(T;x)]^2 \frac{dx}{\sqrt{1-x^2}} = \pi,$$

which should be compared to Exercise 1.4.9.

1.5.22.   Generalization of Exercise 1.5.20. Prove that if k is *even* and $n(1), \ldots, n(k)$ are distinct integers satisfying $1 \leqslant n(i) \leqslant n$ then

$$\int_{-1}^{1} l_{n(1),n}(T;x) l_{n(2),n}(T;x) \cdots l_{n(k),n}(T;x) \frac{dx}{\sqrt{1-x^2}} = 0.$$

Hint. $l_{n(1)}(x) \cdots l_{n(k)}(x) = c(T_n(x))^{k-1} T_n(x)/[(x - \xi_{n(1)}) \cdots (x - \xi_{n(k)})]$ and recall Exercise 1.2.12.

1.5.23.   If $n(1), \ldots, n(m)$ are distinct integers that satisfy $1 \leqslant n(i) \leqslant n$, show that

$$\int_{-1}^{1} \mathfrak{h}_{n(1)}(x) \cdots \mathfrak{h}_{n(m)}(x) \frac{dx}{\sqrt{1-x^2}} = 0,$$

where the $\mathfrak{h}_{n(i)}$ are defined in (1.72). However, also show that

$$\int_{-1}^{1} h_i(x) h_k(x) \frac{dx}{\sqrt{1-x^2}} \neq 0,$$

where the h_j are defined in (1.71).

1.5.24.   Suppose $f(x)$ is continuous on I and W_{2n-1} is the Hermite interpolating polynomial defined by

$$W_{2n-1}(\xi_j) = f(\xi_j), \qquad j = 1, \ldots, n,$$
$$W'_{2n-1}(\xi_j) = y'_j, \qquad j = 1, \ldots, n,$$

where the $y'_j, j = 1, \ldots, n$, are any given real numbers. Show that

$$\lim_{n \to \infty} \int_{-1}^{1} W_{2n-1}(x) \frac{dx}{\sqrt{1-x^2}} = \int_{-1}^{1} \frac{f(x)}{\sqrt{1-x^2}} dx.$$

1.5.25.   If $F(\theta) = f(\cos \pi\theta) = f(x)$ satisfies $F''(\theta) \geqslant 0$ and $F'''(\theta) \geqslant 0$ for $0 \leqslant \theta \leqslant 1$, show that the Gauss–Chebyshev quadrature formulas increase monotonically to (1.114) as n increases.

Any $p \in \mathscr{P}_n$ has a "Chebyshev expansion," i.e., it can be written as

$$p(x) = \frac{A_0}{2} + A_1 T_1(x) + \cdots + A_n T_n(x). \tag{1.126}$$

The coefficients $A_0, A_1, \ldots, A_n$ are easily determined. We multiply both sides of (1.126) by $T_m(x)(1-x^2)^{-1/2}$ and integrate the resulting equality to obtain

$$\int_{-1}^{1} p(x) T_m(x) \frac{dx}{\sqrt{1-x^2}} = \sum_{k=0}^{n}{}' A_k \int_{-1}^{1} T_k(x) T_m(x) \frac{dx}{\sqrt{1-x^2}}.†$$

The orthogonality relationship (1.89) now yields

$$A_k = \frac{2}{\pi} \int_{-1}^{1} p(x) T_k(x) \frac{dx}{\sqrt{1-x^2}}, \quad k = 0, \ldots, n, \tag{1.127}$$

as the formula for the coefficients in (1.126). (It was to avoid singling out the case $k = 0$ that we took the first term in (1.126) to be $A_0/2$.) Let us, for example, obtain the Chebyshev expansion (1.126), for $L_{n-1}(f, T; x) \in \mathscr{P}_{n-1}$. Suppose that

$$L_{n-1}(f, T; x) = \sum_{m=0}^{n-1}{}' \alpha_m T_m(x), \tag{1.128}$$

---

†The notation Σ' is defined following (1.102).

then

$$\alpha_m = \frac{2}{\pi} \int\limits_{-1}^{1} L_{n-1}(x)\, T_m(x)\, \frac{dx}{\sqrt{1-x^2}} \; ;$$

but $L_{n-1}(x)\, T_m(x) \in \mathscr{P}_{2n-1}$, hence (1.118) applies and we obtain

$$\alpha_m = \frac{2}{n} \sum_{j=1}^{n} L_{n-1}(\xi_j)\, T_m(\xi_j)$$

or

$$\alpha_m = \frac{2}{n} \sum_{j=1}^{n} f(\xi_j)\, T_m(\xi_j), \qquad m = 0, \ldots, n-1. \tag{1.129}$$

In particular, if $f(\xi_k) = 1$ and $f(\xi_i) = 0$, $i \neq k$, then $L_{n-1}(f, T; x) = l_{k,n}(T; x)$, and we obtain

$$l_{k,n}(T; x) = l_k(x) = \frac{2}{n} \sum_{m=0}^{n-1}{}' T_m(\xi_k)\, T_m(x). \tag{1.130}$$

The expression (1.130) has some remarkable consequences. In the preceding section we proved that

$$\lim_{n \to \infty} \int\limits_{-1}^{1} L_{n-1}(f, T; x)\, \frac{dx}{\sqrt{1-x^2}} = \int\limits_{-1}^{1} \frac{f(x)}{\sqrt{1-x^2}}\, dx \tag{1.131}$$

for $f(x)$ continuous on I. The integral appearing on the left-hand side of (1.131) is precisely (1.123) and so we had a particularly simple numerical integration formula for (1.114). We wish to show next that

$$\int\limits_{-1}^{1} L_{n-1}(f, T; x)\, dx = \sum_{i=1}^{n} f(\xi_i) \int\limits_{-1}^{1} l_i(x)\, dx \tag{1.132}$$

is an *effective* numerical integration formula for

$$\int\limits_{-1}^{1} f(x)\, dx$$

in the sense that

$$\lim_{n \to \infty} \int_{-1}^{1} L_{n-1}(f, T; x) \, dx = \int_{-1}^{1} f(x) \, dx.$$

THEOREM 1.6. *If*

$$\mu_i = \mu_i^{(n)} = \int_{-1}^{1} l_{i,n}(T; x) \, dx, \tag{1.133}$$

then, if $f(x)$ is continuous on I,

$$\lim_{n \to \infty} \sum_{k=1}^{n} \mu_i^{(n)} f(\xi_j^{(n)}) = \int_{-1}^{1} f(x) \, dx. \tag{1.134}$$

Proof. We first establish that $\mu_i > 0$, $i = 1, \ldots, n$.

$$\mu_i = \int_{0}^{\pi} l_i(\cos \theta) \sin \theta \, d\theta$$

$$= \frac{1}{n} \sum_{m=0}^{n-1}{}' \left(2 \int_{0}^{\pi} \sin \theta \cos m\theta \, d\theta \right) \cos m\theta_i,$$

where we have used the trigonometric form of (1.130). Now

$$2 \int_{0}^{\pi} \sin \theta \cos m\theta \, d\theta = \int_{0}^{\pi} (\sin (m+1)\theta - \sin (m-1)\theta) \, d\theta$$

$$= \begin{cases} 0, & m \text{ odd}, \\ -2\left(\dfrac{1}{m-1} - \dfrac{1}{m+1} \right), & m \text{ even } (m > 0), \\ 4, & m = 0. \end{cases}$$

Therefore

$$\mu_i = \frac{2}{n} \left[1 - (1 - \tfrac{1}{3}) \cos 2\theta_i - (\tfrac{1}{3} - \tfrac{1}{5}) \cos 4\theta_i \right.$$

$$\left. - \cdots - \left(\frac{1}{m-1} - \frac{1}{m+1} \right) \cos m\theta_i - \cdots - \left(\frac{1}{k-1} - \frac{1}{k+1} \right) \cos k\theta_i \right],$$

where $k = n - 1$ for odd n and $k = n - 2$ for even n; but $\cos m\theta_i \leqslant 1$, hence

$$\mu_i \geqslant \frac{2}{n} \left[1 - (1 - \tfrac{1}{3}) - (\tfrac{1}{3} - \tfrac{1}{5}) - \cdots - \left(\frac{1}{m-1} - \frac{1}{m+1} \right) - \cdots \right.$$

$$\left. - \left(\frac{1}{k-1} - \frac{1}{k+1} \right) \right] \geqslant \frac{2}{n} \cdot \frac{1}{k+1} \geqslant \frac{2}{n^2} > 0,$$

and the positivity of μ_i, $i = 1, \ldots, n$, is established.

Suppose now that $p(x)$ is a polynomial, say $p \in \mathscr{P}_k$; then

$$L_n(p, T; x) = p(x), \qquad n \geqslant k$$

and so

$$\int_{-1}^{1} L_n(p, T; x)\, dx = \int_{-1}^{1} p(x)\, dx, \qquad n \geqslant k, \quad k = 1, \ldots.$$

Thus the theorem is proved for polynomial $f(x)$, without recourse to the positivity of the μ_i.

If $f(x)$ is continuous on I, then, according to the Weierstrass approximation theorem (Theorem 1.4), given any $\epsilon > 0$, we can find a polynomial $p(x)$ such that

$$f(x) = p(x) + \delta(x) \tag{1.135}$$

with

$$|\delta(x)| < \epsilon, \qquad x \in I. \tag{1.136}$$

Equation 1.135 implies that

$$\int_{-1}^{1} L_{n-1}(f, T; x)\, dx = \int_{-1}^{1} L_{n-1}(p, T; x)\, dx + \int_{-1}^{1} L_{n-1}(\delta, T; x)\, dx. \tag{1.137}$$

Now

$$\int_{-1}^{1} L_{n-1}(\delta, T; x)\, dx = \sum_{i=1}^{n} \mu_i \delta(\xi_i);$$

hence

$$\left| \int_{-1}^{1} L_{n-1}(\delta, T; x)\, dx \right| \leqslant \sum_{i=1}^{n} |\mu_i|\, |\delta(\xi_i)| < \epsilon \sum_{i=1}^{n} \mu_i,$$

in view of (1.136) and the positivity of the μ_i, but

$$\sum_{i=1}^{n} \mu_i = \sum_{i=1}^{n} \int_{-1}^{1} l_i(x)\, dx = \int_{-1}^{1} \left(\sum_{i=1}^{n} l_i(x) \right) dx = 2,$$

where we make use of Exercise 1.3.1. Therefore

$$-2\epsilon < \int_{-1}^{1} L_{n-1}(\delta, T; x)\, dx < 2\epsilon,$$

whereas integration of (1.135) implies that

$$\int_{-1}^{1} p(x)\, dx = \int_{-1}^{1} f(x)\, dx - \int_{-1}^{1} \delta(x)\, dx,$$

where

$$-2\epsilon < \int_{-1}^{1} \delta(x)\, dx < 2\epsilon.$$

If we now restrict n to be greater than N, the degree of $p(x)$, (1.137) yields

$$\int_{-1}^{1} L_{n-1}(f, T; x)\, dx = \int_{-1}^{1} p(x)\, dx + \int_{-1}^{1} L_{n-1}(\delta, T; x)\, dx$$

$$= \int_{-1}^{1} f(x)\, dx - \int_{-1}^{1} \delta(x)\, dx + \int_{-1}^{1} L_{n-1}(\delta, T; x)\, dx.$$

Thus for $n > N(\epsilon)$

$$\left| \int_{-1}^{1} L_{n-1}(f, T; x)\, dx - \int_{-1}^{1} f(x)\, dx \right| < 4\epsilon.$$

The theorem is proved.   ■

Remark.   Fejér [4] proves this theorem in the more general case that f is Riemann integrable on I.

Although the sequence of interpolating polynomials in the zeros of the Chebyshev polynomial does not converge uniformly to every continuous function, we conclude this chapter on a positive note by showing that this sequence does converge in the mean. As a byproduct we thus obtain another proof of Theorem 1.6.

THEOREM 1.7.

If f is continuous on I, then

$$\lim_{n \to \infty} \int_{-1}^{1} [f(x) - L_n(f, T; x)]^2 \frac{dx}{\sqrt{1 - x^2}} = 0.$$

Proof. As in the proof of Theorem 1.6, given ϵ satisfying $0 < \epsilon < 1$, let $p(x) \in \mathscr{P}_N$ be a polynomial satisfying

$$|f(x) - p(x)| < \epsilon, \qquad x \in I;$$

then

$$\int_{-1}^{1} [f(x) - p(x)]^2 \frac{dx}{\sqrt{1 - x^2}} < \epsilon^2 \pi. \tag{1.138}$$

If $n > N$, then $L_n(p, T; x) = p(x)$; hence

$$\int_{-1}^{1} [L_n(f, T; x) - p(x)]^2 \frac{dx}{\sqrt{1 - x^2}} = \int_{-1}^{1} [L_n(f - p, T; x)]^2 \frac{dx}{\sqrt{1 - x^2}}$$

$$= \int_{-1}^{1} \left[\sum_{i=1}^{n+1} \{f(\xi_i^{(n+1)}) - p(\xi_i^{(n+1)})\} l_{i, n+1}(T; x) \right]^2 \frac{dx}{\sqrt{1 - x^2}}$$

$$= \sum_{i=1}^{n+1} \frac{\pi}{n+1} \{f(\xi_i^{(n+1)}) - p(\xi_i^{(n+1)})\}^2 < \pi\epsilon^2, \tag{1.139}$$

where we have used Exercises 1.5.20 and 1.5.21. The theorem now follows from (1.138) and (1.139), in view of the inequality $(A - B)^2 \leqslant 2(A^2 + B^2)$, which, in turn, is a consequence of the identity $(A - B)^2 + (A + B)^2 = 2(A^2 + B^2)$. ∎

Remark 1. Since $(1 - x^2)^{1/2} \leqslant 1$ on I, Theorem 1.7 implies that

$$\lim_{n \to \infty} \int_{-1}^{1} [f(x) - L_n(f, T; x)]^2 \, dx = 0, \tag{1.140}$$

and Theorem 1.6 follows by Schwarz's inequality. An application of the same inequality shows that (1.131) is also a consequence of Theorem 1.7.

Remark 2. Much more than Theorem 1.7 is known to be true. Erdös and Feldheim (cf. Feldheim [1]) have shown that

$$\lim_{n\to\infty} \int_{-1}^{1} |f(x) - L_n(f, T; x)|^p \frac{dx}{\sqrt{1-x^2}} = 0$$

for all $p > 0$. The theory of mean convergence with respect to other sets of nodes and weight functions is discussed in Askey [2].

EXERCISES 1.5 (continued)

1.5.26. It is quite remarkable that the Chebyshev polynomials also enjoy orthogonality properties on finite point sets in I; for example, show that if $\xi_1, \ldots, \xi_n$ are the zeros of $T_n(x)$ then

$$\sum_{j=1}^{n} T_k(\xi_j) T_m(\xi_j) = \begin{cases} \dfrac{(-1)^p + (-1)^q}{2} n, & \text{if } \begin{cases} k + m = 2pn \\ \text{and} \\ |k - m| = 2qn, \end{cases} \\ (-1)^s \dfrac{n}{2}, & \text{if } \begin{cases} k + m = 2sn \\ \text{and} \\ |k - m| \neq 2rn \end{cases} \\ & \text{or } \begin{cases} |k - m| = 2sn \\ \text{and} \\ k + m \neq 2rn, \end{cases} \\ 0, & \text{otherwise.} \end{cases} \tag{1.141}$$

1.5.27. Show that the coefficients in the Chebyshev expansion of $p \in \mathscr{P}_{n-1}$,

$$p(x) = \frac{A_0}{2} + A_1 T_1(x) + \cdots + A_{n-1} T_{n-1}(x), \tag{1.142}$$

can be obtained by the formula

$$A_m = \frac{2}{n} \sum_{j=1}^{n} p(\xi_j) T_m(\xi_j), \qquad m = 0, 1, \ldots, n - 1. \tag{1.143}$$

1.5.28.   Show that

$$\sum_{j=0}^{n}{}'' T_k(\eta_j)\, T_m(\eta_j) = \begin{cases} 0, & \begin{array}{l} k+m \neq 2pn \\ \text{and} \\ |k-m| \neq 2qn, \end{array} \\[1em] n, & \begin{array}{l} k+m = 2pn \\ \text{and} \\ |k-m| = 2qn, \end{array} \\[1em] \dfrac{n}{2}, & \begin{array}{l} k+m = 2pn \\ \text{and} \\ |k-m| \neq 2qn \\ \text{or} \\ k+m \neq 2pn \\ \text{and} \\ |k-m| = 2qn, \end{array} \end{cases} \tag{1.144}†$$

where $\eta_0, \ldots, \eta_n$ are the extrema of $T_n(x)$.

Formula 1.144 is another "orthogonality" property of the Chebyshev polynomials.

Hint. $\cos n\theta = \mathrm{Re}\,(e^{in\theta})$.

1.5.29.   Show that

$$\int_{-1}^{1} f(x)\, \frac{dx}{\sqrt{1-x^2}} = \frac{\pi}{n} \sum_{i=0}^{n}{}'' f(\eta_i^{(n)}), \qquad f \in \mathscr{P}_{2n-1}. \tag{1.145}$$

Hint.   Verify (1.145) for $T_0, T_1, \ldots, T_{2n-1}$. The formula is variously named after Lobatto and Markov.

1.5.30.   Show that there is no quadrature formula

$$\int_{-1}^{1} f(x)\, \frac{dx}{\sqrt{1-x^2}} = c_0 f(-1) + \sum_{i=1}^{n-1} c_i f(x_i) + c_n f(1) \tag{1.146}$$

exact for $f \in \mathscr{P}_m$ with $m \geqslant 2n$, and (1.145) is the only formula of this type exact for $f \in \mathscr{P}_{2n-1}$.

Hint.   Show that (1.145) cannot hold for $f \in \mathscr{P}_{2n}$, and then show that if (1.146) holds it must coincide with (1.145).

---

† The notation Σ'' is defined in (1.99).

1.5.31. If

$$x^n = \sum_{j=0}^{n}{}' B_j^{(n)} T_j(x),$$
(1.147)

show that

$$B_{n-2k}^{(n)} = 2^{1-n}\binom{n}{k}, \qquad k = 0, \ldots, \left[\frac{n}{2}\right],$$
(1.148)

and that

$$B_j^{(n)} = 0, \qquad j \neq n - 2k.$$

1.5.32. If

$$\sum_{k=0}^{n} a_k x^k = \sum_{k=0}^{n}{}' A_k T_k(x),$$

then

$$A_k = \frac{1}{2^{k-1}}\left[a_k + \sum_{j=1}^{[(n-k)/2]} \frac{\binom{k+2j}{j} a_{k+2j}}{2^{2j}} \right].$$

1.5.33. If

$$T_n'(x) = \sum_{j=0}^{n-1} A_j T_j(x),$$

show that $A_j \geqslant 0$. For which j is $A_j = 0$?

Hint. See Exercises 1.2.13 and 1.2.15c.

1.5.34. If

$$T_n^{(k)}(x) = \sum_{j=0}^{n-k} A_{jk} T_j(x), \qquad k = 0, 1, 2, \ldots, n,$$

show that

$$A_{jk} \geqslant 0, \qquad k = 0, 1, \ldots, n; \; j = 0, 1, \ldots, n - k.$$

For which j is $A_{jk} = 0$?

Hint. Use mathematical induction on k and Exercise 1.5.33.

1.5.35. Show that

$$|T_n^{(k)}(x)| \leqslant T_n^{(k)}(1), \qquad x \in I, \; k = 0, 1, \ldots, n,$$

with equality only for $x = \pm 1$, $k \geqslant 1$, thus generalizing Exercise 1.2.4, in view of Exercise 1.5.6.

Hint. This result follows immediately from Exercise 1.5.34.

1.5.36. (Feldheim [1]) Show that

$$[l_{k,n}(T; \cos \theta)]^2 = \frac{2}{n} \sum_{m=0}^{2n-2}{}' \cos m\theta_k \cos m\theta + \frac{1}{n^2} \sum_{m=2}^{2n-2} c_{m,k} \cos m\theta, \tag{1.149}$$

where

$$c_{m,k} = \frac{\sin m\theta_k \cos \theta_k - m \cos m\theta_k \sin \theta_k}{\sin \theta_k}, \qquad k = 1, \ldots, n.$$

Hint. Use (1.130).

1.5.37. Show that

$$\sum_{k=1}^{n} [l_{k,n}(T; \cos \theta)]^2 = \begin{cases} 1 - \dfrac{1}{2n} + \dfrac{\sin (2n-1)\theta}{2n \sin \theta}, & 0 < \theta < \pi, \\[2ex] 2 - \dfrac{1}{n}, & \theta = 0, \pi. \end{cases}$$

Hint. Sum (1.149) and recall Exercises 1.2.13 and 1.5.26.

1.5.28. Show that

$$\lim_{n \to \infty} \sum_{k=1}^{n} [l_{k,n}(T; x)]^2 = \begin{cases} 1, & -1 < x < 1, \\ 2, & x = \pm 1. \end{cases}$$

[Compare this result with Exercise 1.4.9. It is interesting to remark here, that although the problem of finding an array of nodes X, such that

$$\max_{-1 \leqslant x \leqslant 1} \sum_{k=1}^{n} |l_{k,n}(X; x)|$$

is minimum is unsolved, the analogous problem of finding an array of nodes such that

$$\max_{-1 \leqslant x \leqslant 1} \sum_{k=1}^{n} [l_{k,n}(X; x)]^2 \tag{1.150}$$

is minimum has been solved. (Fejér [3].) The array of nodes that produces the minimum of (1.150) consists of the zeros of

$$F_n(x) = \int_x^1 P_{n-1}(t) \, dt,$$

where $\{P_n(x)\}$ are the Legendre polynomials, i.e., the ultraspherical polynomials with $\lambda = \frac{1}{2}$. Remarkably enough the value of the minimum of (1.150) is 1.]

1.5.39. Show that for each integer $m \geqslant 2$ and $x \in I$

$$\left(\sum_{k=1}^{n} |\, l_{k,n}(T;x)\,|^m \right)^{1/m} \leqslant \sqrt{2}.$$

Hint. First consider $m = 2r$ and observe that, in notation suggested by Exercise 1.4.2,

$$1 = \left[\sum_{k=1}^{n} v_k(T)\, l_k^2 \right]^r \geqslant \sum_{k=1}^{n} v_k^r \, l_k^{2r}.$$

Then use Schwarz's inequality for the odd m. Compare with Exercise 1.4.9.

1.5.40. Show that

$$\max_{-1 \leqslant x \leqslant 1} |\, l_{1,n}(T;x)\,| = |\, l_{1,n}(T;1)\,|$$

and the sequence $|\, l_{1,n}(T;1)\,|$ is monotone increasing, with n, to $4/\pi$.

1.5.41. If $n \geqslant 3$ and

$$\max_{-1 \leqslant x \leqslant 1} |\, l_{j,n}(T;x)\,| = |\, l_{j,n}(T;u_j)\,|, \qquad j = 2, \ldots, n-1,$$

then, if $u_j = \cos \alpha_j$, we have [recalling the notation of (1.16)]

$$|\alpha_j - \theta_j^{(n)}| < \frac{\pi}{2n}.$$

Hint. Apply the M. Riesz theorem (Exercise 1.3.20) to $l_{j,n}(T; \cos \theta) \in \mathcal{T}_{n-1}$.

1.5.42. With the same hypothesis and notation as in Exercise 1.5.41 show that for $j = 2, \ldots, n-1$,

$$\frac{1 - \xi_j u_j}{1 - \xi_j^2} \geqslant \frac{13}{18}.$$

Hint. If $\theta_j^{(n)} = 3\mu$, then $\theta_j^{(n)} - \pi/(2n) \geqslant 2\mu$; hence

$$A = \frac{\cos(\theta_j^{(n)} - \pi/2n) - \cos \theta_j^{(n)}}{1 - \cos \theta_j^{(n)}} \leqslant \frac{\cos 2\mu - \cos 3\mu}{1 - \cos 3\mu} \leqslant \frac{5}{9},$$

so that

$$A \frac{\cos \theta_j^{(n)}}{1 + \cos \theta_j^{(n)}} \leqslant \frac{5}{18}.$$

1.5.43. Show that for $j = 2, \ldots, n - 1$, and $n \geqslant 3$

$$(l_{j,n}(T; u_j))^2 < \frac{18}{13} < \left(\frac{1 + \sqrt{2}}{2} \right) = (l_{1,2}(T; 1))^2.$$

Hint. Recall Exercise 1.4.1.

1.5.44. (Erdös and Grünwald [1])   Show that for $n \geqslant 1$

$$\sup_n \max_j \max_{-1 \leqslant x \leqslant 1} |l_{j,n}(T; x)| = \frac{4}{\pi}.$$

Hint. Put together the four preceding exercises. Compare with Exercise 1.4.10.

1.5.45. Prove that

$$T_n(x) = \begin{vmatrix} 2x & -1 & 0 & \ldots & & 0 & 0 \\ -1 & 2x & -1 & \ldots & & 0 & 0 \\ 0 & -1 & 2x & -1 & \ldots & 0 & 0 \\ \vdots & \vdots & \vdots & \vdots & & \vdots & \vdots \\ 0 & 0 & 0 & \ldots & -1 & 2x & -1 \\ 0 & 0 & 0 & \ldots & 0 & -1 & x \end{vmatrix},$$

where the determinant is $n \times n$. Also show that the corner elements x and $2x$ can be interchanged.

Hint. Expand in terms of elements of the first column and use (1.101).

1.5.46. (Korsak and Schubert [1])   Prove that

$$D = \begin{vmatrix} 1 & -2x & 1 & \ldots & & & & 0 & 0 \\ 0 & 1 & -2x & 1 & \ldots & & & 0 & 0 \\ 0 & 0 & 1 & -2x & 1 & \ldots & & 0 & 0 \\ \vdots & \vdots & \vdots & \vdots & \vdots & \vdots & \vdots & & \vdots \\ 0 & 0 & 0 & \ldots & & 1 & -2x & 1 \\ 1 & 0 & 0 & \ldots & & 0 & 1 & -2x \\ -2x & 1 & 0 & \ldots & & & 0 & 1 \end{vmatrix} = 2(1 - T_n(x)),$$

where D is $n \times n$.

Hint. Show that D is zero at $x = \eta_k$, $k = 0, 2, 4, \ldots$, by establishing that

$$T_0(\eta_k)C_1 + \cdots + T_{n-1}(\eta_k)C_n = 0,$$

where C_j is the jth column vector of the matrix of D. Then show that if $x \in I$ and

$$E = \begin{vmatrix} 1 & -\tau & 0 & \cdots & & & 0 \\ 0 & 1 & -\tau & \cdots & & & 0 \\ 0 & 0 & 1 & -\tau & & \cdots & 0 \\ \vdots & \vdots & \vdots & \vdots & \vdots & \vdots & \vdots \\ 0 & \cdots & & & 0 & 1 & -\tau \\ -\tau & \cdots & & & 0 & 0 & 1 \end{vmatrix},$$

where $\tau = x - \sqrt{x^2 - 1}$, then

$$D = \bar{E} \cdot E \geqslant 0,$$

where $\bar{E}$ is the complex conjugate of E.

2

EXTREMAL PROPERTIES

One of the most remarkable properties of the Chebyshev polynomial, $T_n(x)$, is that $\tilde{T}_n(x)$ (the Chebyshev polynomial normalized so that its leading coefficient is 1) has the smallest maximum absolute value on I: $[-1, 1]$ among all $p(x) = x^n + a_{n-1}x^{n-1} + \cdots + a_0$ [cf. (1.109)] (This property is one basis for the wide utility of the Chebyshev polynomials in numerical analysis, a topic to which we turn in Chapter 3.) Let us begin by proving this fact. We recall that if $g(x)$ is continuous on I

$$\|g\| = \max_{-1 \leqslant x \leqslant 1} |g(x)|.$$

THEOREM 2.1

If $p(x) = x^n + a_{n-1}x^{n-1} + \cdots + a_0$, *then*

$$\|p\| \geqslant \|\tilde{T}_n\| = \begin{cases} 2^{1-n}, & n > 0, \\ 1, & n = 0, \end{cases}$$

with equality only if $p = \tilde{T}_n$.

Proof. Suppose that $n > 0$ and $\|p\| \leqslant \|\tilde{T}_n\| = 2^{1-n}$, then $|p(x)| \leqslant 2^{1-n}$ throughout I and $\tilde{T}_n(\eta_k) = (-1)^k 2^{1-n}$, $k = 0, \ldots, n$ [η_k are as defined in (1.19)]. Suppose that $p \neq \tilde{T}_n$, then $q = \tilde{T}_n - p \in \mathscr{P}_{n-1}$ and is not identically zero. Suppose that $q(\eta_0) = \cdots = q(\eta_{k-1}) = 0$, $q(\eta_k) \neq 0$, and $q(\eta_m) \neq 0$ but $q(\eta_{m+1}) = \cdots = q(\eta_n) = 0$. Since q is not the zero polynomial, $k \leqslant n$ and $m \geqslant 0$. Note now that if $q(\eta_i) \neq 0$, then sgn $q(\eta_i) = $ sgn $\tilde{T}_n(\eta_i) = (-1)^i$. Suppose that $q(\eta_i) \neq 0$, $q(\eta_{i+1}) = \cdots = q(\eta_{i+(j-1)}) = 0$, $q(\eta_{i+j}) \neq 0$. Then certainly q has at least $j - 1$ zeros in $[\eta_i, \eta_{i+j}]$, but if j is even sgn $q(\eta_i) = $ sgn $q(\eta_{i+j})$ and so q has an even number of zeros (counting multiple zeros as many times as their multiplicity) in $[\eta_i, \eta_{i+j}]$, hence at least j zeros. If j is odd, sgn $q(\eta_i) = -$sgn $q(\eta_{i+j})$ and so q has an odd number of zeros (counting multiple zeros as many times as their multiplicity) in $[\eta_i, \eta_{i+j}]$, hence at least j zeros. Thus q has at least j zeros in $[\eta_i, \eta_{i+j}]$, hence at least $m - k$ zeros in $[\eta_k, \eta_m]$. But q also has k zeros in

$[\eta_0, \eta_k]$ and $n - m$ zeros in $[\eta_m, \eta_n]$, making a total of $(m - k) + k + (n - m) = n$ zeros in I. Since $q \in \mathscr{P}_{n-1}, q = 0$. Thus either $\| p \| > \| \tilde{T}_n \|$ or $p = \tilde{T}_n$. ∎

COROLLARY 2.1.1

If $p(x) = a_0 + a_1 x + \cdots + a_k x^k, a_k \neq 0$, and $p \neq a_k \tilde{T}_k$, then there exist $x_0 \in I$ such that

$$| p(x_0) | > \frac{|a_k|}{2^{k-1}}.$$

Theorem 2.1 has two interesting reinterpretations:

A.   The polynomial in $\mathscr{P}_{n-1}$ closest to the function $f(x) = x^n$, where closeness is measured by $\| f - p \|, p \in \mathscr{P}_{n-1}$, is

$$p^* = x^n - \tilde{T}_n.$$

B.   Among all $p \in \mathscr{P}_n$ satisfying $\| p \| = 1$ the largest value of

$$| Fp | = \left| \frac{p^{(n)}(0)}{n!} \right|$$

is 2^{n-1} and this value is assumed only for $p = \pm T_n$.

(A) is an example of a rich mathematical area, the uniform approximation of functions by polynomials, and (B) is an example of the problem of maximizing a linear functional on the space $\mathscr{P}_n$. To see how the Chebyshev polynomials fit into these larger schemes we next explore both areas in some detail.

A.   UNIFORM APPROXIMATION OF CONTINUOUS FUNCTIONS

2.1.   Convex Sets in n-Space

Our investigation of uniform approximation will be quite wide-ranging on its circuitous route back to the Chebyshev polynomials. For this purpose we need some preliminary material about convex sets in (real) n-dimensional space, $\mathbb{R}^n$.

DEFINITION 2.1

A set, S in $\mathbb{R}^n$ is convex if $s_1, s_2, \in S$ implies $\lambda s_1 + (1 - \lambda)s_2 \in S$ for $0 \leqslant \lambda \leqslant 1$; that is, together with any two points S contains the line segment joining them.

DEFINITION 2.2

If nonnegative numbers $\lambda_1, \ldots, \lambda_m$ *satisfy*

$$\sum_{i=1}^{m} \lambda_i = 1$$

and $y_1, \ldots, y_m \in \mathbb{R}^n$,

$$\sum_{i=1}^{m} \lambda_i y_i$$

is called a convex combination of $y_1, \ldots, y_m$.

DEFINITION 2.3

Given a set S, form the set $\hat{S}$, *of all convex combinations of points of S.* $\hat{S}$ *is called the convex hull of S.*

THEOREM 2.2 (Carathéodory)

Let S be a subset of $\mathbb{R}^n$ *and y a point of* $\hat{S}$. *Then there exists a subset* S_1 *of S, containing at most n + 1 points, such that* $y \in \hat{S}_1$. *Indeed, given any* $z \in \mathbb{R}^n$, *we may choose* S_1 *to consist of z and at most n points of S.*

Proof. If we can prove this result for $z = 0$, namely, for any subset of $\mathbb{R}^n$, B, and $w \in \hat{B}$,

$$w = \sum_{i=1}^{r} \mu_i b_i, \qquad \mu_i \geqslant 0, \qquad \sum_{i=1}^{r} \mu_i \leqslant 1,$$

where $r \leqslant n$ and $b_i \in B$, then it is true for arbitrary z in n-space. For given S and $y \in \hat{S}$ let $B = \{x \in \mathbb{R}^n \mid x = s - z, s \in S\}$. Then $w = y - z \in \hat{B}$; hence

$$y - z = \sum_{i=1}^{r} \mu_i b_i = \sum_{i=1}^{r} \mu_i (s_i - z), \qquad s_i \in S.$$

Thus

$$y = \left(1 - \sum_{i=1}^{r} \mu_i\right) z + \sum_{i=1}^{r} \mu_i s_i.$$

Suppose then that $z = 0$. If $y = 0$, $y = z$ and the theorem is proved. Suppose that $y \neq 0$. Since $y \in \hat{S}$, there exist $s_1, \ldots, s_r \in S$ and *positive* $\lambda_1, \ldots, \lambda_r$ which satisfy

$$\sum_{i=1}^{r} \lambda_i \leqslant 1 \qquad\qquad\qquad\qquad (2.1)$$

such that

$$y = \sum_{i=1}^{r} \lambda_i s_i. \tag{2.2}$$

We show that if r is chosen to be the least integer such that (2.1) and (2.2) hold then $r \leqslant n$. Suppose that $r > n$; then the s_i are linearly dependent and we can find real numbers c_i such that

$$\sum_{i=1}^{r} c_i \geqslant 0$$

(in particular, at least one c_i is positive) and

$$\sum_{i=1}^{r} c_i s_i = 0.$$

If t is any real number, we have

$$y = \sum_{i=1}^{r} (\lambda_i - tc_i) s_i.$$

The numbers $t_i(t) = \lambda_i - tc_i$ are all positive for $t = 0$, hence for positive and sufficiently small t. Since at least one c_i is positive, there is a smallest positive value of t, say t', such that at least one of the $t_i(t)$ vanishes. [Note that none of the $t_i(t')$ is negative and at least one of them is positive, since $y \neq 0$.] Thus

$$y = \sum_{i=1}^{r} t_i(t') s_i$$

and at least one of $t_i(t') = 0$, but

$$\sum_{i=1}^{r} t_i(t') = \sum_{i=1}^{r} \lambda_i - t' \sum_{i=1}^{r} c_i \leqslant 1,$$

which contradicts the assumed minimal property of r. ■

DEFINITION 2.4

If $x: (x_1, \ldots, x_n)$ and $y: (y_1, \ldots, y_n)$ are two points of $\mathbb{R}^n$, $(x, y) = x_1 y_1 + \cdots + x_n y_n$ is called the inner product of x and y; $d(x, y) = (x - y, x - y)^{1/2}$ is the distance between x and y.

THEOREM 2.3 (Separating Hyperplane Theorem)

Let C be a closed convex set not containing the origin in $\mathbb{R}^n$. Then there exists *a hyperplane, H, defined by* $h(x) = (x, a) + b = 0$ *such that* $h(0) < 0$ *and* $h(x) > 0$ *for all* $x \in C$; *that is, H strictly separates C from 0.*

Proof.   Since C is closed, there exists a point of C, x^* which is closest to the origin. Take

$$h(x) = \left(x - \frac{x^*}{2}, x^*\right).$$

Then $h(0) = -\frac{1}{2}(x^*, x^*) < 0$ and $h(x^*) = \frac{1}{2}(x^*, x^*) > 0$.

We claim that $h(x) > 0$ for all $x \in C$. Suppose it is not, that $x' \in C$, and $h(x') \leqslant 0$. Then

$$(x', x^*) \leqslant \tfrac{1}{2}(x^*, x^*). \tag{2.3}$$

Since x' and $x^* \in C$, every point $\lambda x' + (1 - \lambda)x^*, 0 \leqslant \lambda \leqslant 1$, is in C, but if

$$f(\lambda) = (\lambda x' + (1 - \lambda)x^*, \lambda x' + (1 - \lambda)x^*)$$
$$= \lambda^2(x', x') + 2\lambda(1 - \lambda)(x', x^*) + (1 - \lambda)^2(x^*, x^*)$$

we have $f(0) = (x^*, x^*)$ and $f'(0) = 2(x', x^*) - 2(x^*, x^*) < 0$, in view of (2.3). Thus there is a point of C closer to 0 than x^*, contradicting the definition of x^*.   ∎

EXERCISES 2.1

2.1.1.   Show that C is convex if, and only if, every convex combination of points of C is contained in C.

2.1.2.   If $S \subset \mathbb{R}^n$, show that $\hat{S}$ is convex.

2.1.3.   Show that $S \subset \hat{S}$.

2.1.4.   Show that $\hat{S}$ is the smallest convex set containing S.

2.1.5.   Show that if S is compact then $\hat{S}$ is also.

2.2.   Characterization of Best Approximations

We are now in a position to discuss a quite general problem of uniform approximation. The setting is as follows. Let B be a compact set in m-space and let $C(B)$ be the set of (real or complex-valued) continuous functions on B, equipped with the uniform (or Chebyshev) norm; that is, if $g \in C(B)$,

$$\|g\| = \max_{y \in B} |g(y)|.$$

Associated with each $g \in C(B)$ is its (nonempty) set of *critical points,*

$$E(g;B) = \{ y \in B / | \, g(y) \, | = \| \, g \, \| \}.$$

Note that $E(g;B)$ is closed, hence compact.

Given V, a k-dimensional subspace of $C(B)$, our objective is to characterize the elements of V that are closest to a given $f \in C(B)$, closeness being measured by the uniform norm. The entire theoretical foundation of our endeavors is contained in the following result.

THEOREM 2.4

Suppose $g \in C(B)$ and $\phi_1, \ldots, \phi_k$ is any basis for V. Let K be the set in k-space described by $(\overline{g(y)} \, \phi_1(y), \ldots, \overline{g(y)} \, \phi_k(y))$ as y runs through $E(g;B)$. Then

$$\| g + v \| \geqslant \| g \|, \qquad \text{all } v \in V \tag{2.4}$$

if, and only if, the origin in k-space is in the convex hull of some subset of r points of K, where $r \leqslant k + 1$ in the real case (i.e., when $C(B)$ consists of real-valued functions) and $r \leqslant 2k + 1$ points in the complex case.

Proof. (i) Suppose that $y_1, \ldots, y_r$ are points of $E(g;B)$ such that

$$0 = \sum_{i=1}^{r} \lambda_i (\overline{g(y_i)} \, \phi_1(y_i), \ldots, \overline{g(y_i)} \, \phi_k(y_i)),$$

where $\lambda_i \geqslant 0$ and $\sum_{i=1}^{r} \lambda_i = 1$. Then

$$0 = \sum_{i-1}^{r} \lambda_i \, \overline{g(y_i)} \, \phi_j(y_i), \qquad j = 1, \ldots, k;$$

hence, for any $v \in V$,

$$0 = \sum_{i=1}^{r} \lambda_i \, \overline{g(y_i)} \, v(y_i)$$

and

$$0 = \mathrm{Re} \sum_{i=1}^{r} \lambda_i \overline{g(y_i)} \, v(y_i) = \sum_{i=1}^{r} \lambda_i \, \mathrm{Re} \, \overline{g(y_i)} \, v(y_i),$$

so that for each $v \in V$ there exists $i(v)$ such that

$$\mathrm{Re} \, \overline{g(y_i)} \, v(y_i) \geqslant 0. \tag{2.5}$$

Given any v, fix i so that (2.5) holds; then

$$\| g + v \|^2 \geqslant | g(y_i) + v(y_i) |^2 = [g(y_i) + v(y_i)] \cdot [\overline{g(y_i)} + \overline{v(y_i)}]$$
$$\geqslant \| g \|^2 + | v(y_i) |^2 + 2 \, \mathrm{Re} \, \overline{g(y_i)} \, v(y_i) \geqslant \| g \|^2,$$

thus establishing (2.4).

(ii) Suppose that (2.4) holds and $\|g\| = 1$. (If $\|g\| = 0$, the theorem is trivial. If $0 < \|g\| \neq 1$, continue the proof with $g_0 = g/\|g\|$, with no loss in generality.)

If $\hat{K}$ is the convex hull of K, we show that $0 \in \hat{K}$. Suppose that $0 \notin \hat{K}$. Since K is compact, so is $\hat{K}$ and there exists a hyperplane that separates 0 from $\hat{K}$ according to Theorem 2.3; that is, there exist complex numbers $c_0, \ldots, c_k$ $(c_0 \neq 0)$ such that the half-space

$$\mathrm{Re}\,(c_0 + c_1 z_1 + \cdots + c_k z_k) \geqslant 0$$

contains $\hat{K}$ but $-\tau = \mathrm{Re}\,c_0 < 0$; that is, for all $y \in E(g;B)$

$$\mathrm{Re}\left(\sum_{j=1}^{k} c_j \overline{g(y)}\, \phi_j(y) \right) \geqslant \tau > 0,$$

or, putting $v_0 = \Sigma_{j=1}^{k}\, c_j \phi_j$,

$$\mathrm{Re}\,\overline{g(y)}\, v_0(y) \geqslant \tau > 0, \qquad y \in E(g;B).$$

Let U_1 be an open set of m-space such that $E(g;B) \subset U_1 \cap B = U$, $\mathrm{Re}\,\bar{g} v_0 \geqslant \tau/2$ on U, and $|g(y)| \leqslant 1 - \delta\,(\delta > 0)$ on $B \setminus U$, which is closed. Now choose $\epsilon > 0$ so that $\epsilon \|v_0\| < \delta$; then

$$\max_{y \in B \setminus U} |g(y) - \epsilon v_0(y)| < 1,$$

but on U

$$|g(y) - \epsilon v_0(y)|^2 = |g(y)|^2 + \epsilon^2 |v_0(y)|^2 - 2\epsilon\, \mathrm{Re}\,\overline{g(y)}\, v_0(y)$$
$$\leqslant 1 + \epsilon^2 \|v_0\|^2 - \epsilon\tau.$$

Since $\epsilon^2 \|v_0\|^2 - \epsilon\tau < 0$ for ϵ sufficiently small,

$$|g(y) - \epsilon v_0(y)| < 1, \qquad y \in U,$$

and putting $v = -\epsilon v_0$, $\|g + v\| < 1 = \|g\|$, contradicting (2.4). Thus $0 \in \hat{K}$ and an application of Carathéodory's theorem (Theorem 2.2) now proves our result. ∎

DEFINITION 2.5

If $f \in C(B)$ is given and $\|f - v^*\| \leqslant \|f - v\|$, all $v \in V$, we call v^* a best approxi-mation to f (on B) out of V, and put

$$\|f - v^*\| = E_V(f).$$

(In case $V = \mathscr{P}_n$ we write $E_n(f)$ for $E_{\mathscr{P}_n}(f)$.)

THEOREM 2.5

v^* is a best approximation to $f \in C(B)$ out of V, if, and only if there exist distinct points $y_1, \ldots, y_r \in E(f - v^*; B)$, and positive numbers $\lambda_1, \ldots, \lambda_r$ such that

$$\sum_{i=1}^{r} \lambda_i \overline{[f(y_i) - v^*(y_i)]} \, v(y_i) = 0, \qquad \text{all } v \in V, \tag{2.6}$$

where $r \leqslant k + 1$ in the real case and $r \leqslant 2k + 1$ in the complex case.

Proof. Replace g by $f - v^*$ in Theorem 2.4. ■

Remark. Theorem 2.5 remains valid if v in (2.6) is replaced by ϕ_j for $j = 1, \ldots, k$.

A useful variant of the characterization of best approximations given in Theorem 2.5 is based on the following considerations.

DEFINITION 2.6

A signature in B, Σ, is a continuous function whose domain is a closed subset of B and whose range is in the unit circle in the complex plane. We call the domain of Σ the base of the signature and denote it by $\sigma(\Sigma)$; Σ' is a subsignature of Σ if Σ' is the restriction of Σ to a subset of $\sigma(\Sigma)$.

DEFINITION 2.7

A signature, $\Sigma[g]$, is said to be associated with $g \in C(B)$ if $\sigma \subseteq E(g; B)$ and $\Sigma(y) - \operatorname{sgn} g(y)$.† (Note that $g = 0$ can have no signature associated with it.)

DEFINITION 2.8

A signature, Σ, is said to be extremal for V if there exist (complex) numbers $\zeta_1, \ldots, \zeta_s$ and distinct points $y_1, \ldots, y_s$ of σ such that

$$\operatorname{sgn} \zeta_i = \overline{\Sigma(y_i)}, \qquad i = 1, \ldots, s \tag{2.7}$$

and

$$\sum_{i=1}^{s} \zeta_i v(y_i) = 0, \qquad \text{all } v \in V; \tag{2.8}$$

$\zeta_1, \ldots, \zeta_s$ are called weights for Σ. An extremal signature, Σ, is called primitive if it has no proper extremal subsignature.

We now obtain an immediate equivalent to Theorem 2.5 in terms of extremal signatures.

† For a complex number $z \neq 0$ $\operatorname{sgn} z = z/|z|$, and $\operatorname{sgn} 0 = 0$ (cf. Exercise 1.3.6).

THEOREM 2.6

A best approximation out of V to $f \notin V$ is v^* if, and only if, there exists an extremal signature for V associated with $f - v^*$ based on r points, where $r \leqslant k + 1$ in the real case and $r \leqslant 2k + 1$ in the complex case.

COROLLARY 2.6.1(Skeleton Theorem)

If v^* is a best approximation to $f \notin V$ on B, it is also a best approximation to f on the base of an extremal signature for V consisting of r points, where $r \leqslant k + 1$ in the real case and $r \leqslant 2k + 1$ in the complex case.

Proof. The base σ of the extremal signature described in Theorem 2.6 works by applying Theorem 2.6, with B replaced by σ. ■

COROLLARY 2.6.2

Let $y_1, \ldots, y_r$ be the base of the extremal signature Σ. Then, if v^* is a best approximation to f out of B and $v \in V$,

$$\|f - v^*\| = E_V(f) \geqslant \min_{1 \leqslant i \leqslant r} \text{Re } \overline{\Sigma(y_i)}[f(y_i) - v(y_i)].$$

Proof. Choose weights for Σ, $\zeta_1, \ldots, \zeta_r$, to satisfy

$$\sum_{i=1}^{r} |\zeta_i| = 1$$

(this can always be done). Then

$$E_V(f) = \sum_{i=1}^{r} |\zeta_i| \, \|f - v^*\| \geqslant \sum_{i=1}^{r} |\zeta_i| \, |f(y_i) - v^*(y_i)|$$

$$\geqslant \left| \sum_{i=1}^{r} \zeta_i [f(y_i) - v^*(y_i)] \right| = \left| \sum_{i=1}^{r} \zeta_i [f(y_i) - v(y_i)] \right|$$

$$= \left| \sum_{i=1}^{r} |\zeta_i| \, \overline{\Sigma(y_i)}[f(y_i) - v(y_i)] \right|$$

$$\geqslant \text{Re} \sum_{i=1}^{r} |\zeta_i| \, \overline{\Sigma(y_i)}[f(y_i) - v(y_i)]$$

$$= \sum_{i=1}^{r} |\zeta_i| \, \text{Re } \overline{\Sigma(y_i)} [f(y_i) - v(y_i)]$$

$$\geqslant \min_{1 \leqslant i \leqslant r} \text{Re } \overline{\Sigma(y_i)} [f(y_i) - v(y_i)]. \quad ■$$

EXERCISES 2.2

2.2.1. **(Kolmogorov Criterion)** Show that v^* is a best approximation to f if, and only if,

$$\min_{y \in E(f-v^*; B)} \text{Re} \quad \overline{\text{sgn}\, [f(y) - v^*(y)]} \cdot v(y) \leqslant 0, \text{ each } v \in V. \tag{2.9}$$

[Thus (2.6) and (2.9) are equivalent.]

Hint. Formula 2.6 implies (2.9). Then show that (2.9) implies that

$$\| f - v \|^2 \geqslant \| f - v^* \|^2.$$

2.2.2. Any signature with an extremal subsignature is itself extremal.

2.2.3. Every extremal signature has a finitely based extremal subsignature.

2.2.4. If Σ is an extremal signature for V, then for each $v \in V$

$$\min_{y \in \sigma(\Sigma)} \text{Re}\, (v(y)\, \overline{\Sigma(y)}) \leqslant 0. \tag{2.10}$$

2.2.5. If (2.10) holds, there is an extremal subsignature of Σ based on r points, where $r \leqslant k + 1$ in the real case and $r \leqslant 2k + 1$ in the complex case.

Hint. Formula 2.10 implies that 0 is a best approximation to Σ on σ out of V.

2.2.6. Every extremal signature has an extremal subsignature based on r points, where $r \leqslant k + 1$ in the real case and $r \leqslant 2k + 1$ in the complex case.

Exercises 2.2.7–2.2.10 deal with the real case.

2.2.7. Points $y_1, \ldots, y_r$ of B are the base of a primitive extremal signature for V if, and only if, every set of $r - 1$ columns of the $k \times r$ matrix

$$\Phi_r = \begin{pmatrix} \phi_1(y_1) \ldots \phi_1(y_r) \\ \vdots \\ \phi_k(y_1) \ldots \phi_k(y_r) \end{pmatrix}$$

is linearly independent and the rank of Φ_r is $r - 1$.

Hint. (i) If, say, the first $r - 1$ columns are linearly dependent, an argument similar to that used in the proof of Theorem 2.2 contradicts the primitivity of the extremal signature. (ii) If the rank is $r - 1$, the columns are linearly dependent.

2.2.8. No proper subset of the base of a primitive extremal signature is the base of an extremal signature.

2.2.9. If Σ is a primitive extremal signature, the only extremal signatures based on $\sigma(\Sigma)$ are $\pm \Sigma$.

2.2.10. If $y_1, \ldots, y_r$ are the base of a primitive extremal signature for V and J

is any set of $r - 1$ distinct indices among $1, \ldots, r$, then, given arbitrary $b_j, j \in J$, there exists $v \in V$ such that

$$v(y_j) = b_j, \qquad j \in J.$$

2.2.11. We have avoided the question whether a best approximation out of V to a given $f \in C(B)$ exists. The answer is that it always does. Fill in the details in the following sketch of a proof of this fact. Show that $\| f - \Sigma_{i=1}^{k} a_i \phi_i \|$, a continuous function of $\mathbf{a}: (a_1, \ldots, a_k)$, assumes its minimum by noting that, since 0 is a possible value of $\mathbf{a}$, we need only consider $\mathbf{a}$ such that $\| \Sigma_{i=1}^{k} a_i \phi_i \| \leqslant 2\| f \|$, and the set of such $\mathbf{a}$ is compact.

2.2.12. Show that $\Sigma(\eta_i) = (-1)^i, i = 0, \ldots, n$, is an extremal signature for $V = \mathscr{P}_{n-1}$ based on the extrema of $T_n(x)$.

Hint. The weights needed to establish (2.7) and (2.8) can be deduced from (1.144) with $0 \leqslant k \leqslant n - 1$ and $m = n$.

2.2.13. Show that Theorem 2.6 implies, in view of Exercise 2.2.12, that a best approximation on I to x^n out of $\mathscr{P}_{n-1}$ is $x^n - \tilde{T}_n$.

2.2.14. (Ehlich and Zeller [3]) Show that a best approximation on the square $B: -1 \leqslant x \leqslant 1, -1 \leqslant y \leqslant 1$ to $x^n y^m$ out of the space

$$V = \left\{ p(x, y) = \sum_{\substack{i=0 \\ i+j<n+m}}^{n} \sum_{j=0}^{m} a_{ij} x^i y^j \right\}$$

is $x^n y^m - \tilde{T}_n(x) \tilde{T}_m(y)$.

Hint. $(\eta_k^{(n)}, \eta_l^{(m)}) \xrightarrow{\Sigma} (-1)^{k+l}$, $\quad k = 0, \ldots, n; \; l = 0, \ldots, m$, is an extremal signature for V.

2.2.15. Let V be the set of linear polynomials in two variables; i.e.,

$$V = \{ax + by + c\} \text{ show that}$$

(i) every primitive extremal signature for V in $\mathbb{R}^2$ is based on at least 3 points and at most 4 points;

(ii) if an extremal signature for V is based on 3 points, they are collinear and the signs alternate;

(iii) if an extremal signature for V is based on 4 points, then either

    (a) they are the vertices of a triangle, all with one sign, and a point inside the triangle with the opposite sign, or

    (b) they are the vertices of a convex quadrilateral with opposite vertices of like sign and adjacent vertices of opposite sign.

2.3. Chebyshev Systems and Uniqueness

Since our ultimate concern is with polynomial approximation, we now further restrict our k-dimensional space of approximators V by requiring that it satisfy the *Chebyshev condition*.

DEFINITION 2.9

V satisfies the Chebyshev condition with respect to B *if, and only if, each $v \in V$ has at most $k - 1$ distinct zeros in B, unless $v = 0$. Any basis of V is said to be a Chebyshev system on B.*

The simplest example of a Chebyshev system is $1, z, \ldots, z^{k-1}$, on any set of the complex plane. A detailed study of the properties of Chebyshev systems is to be found in Karlin and Studden [1].

It is clear that V satisfies the Chebyshev condition (with respect to B) if, and only if, for each set of k distinct points of B, $y_1, \ldots, y_k$, we have

$$\Delta = \begin{vmatrix} \phi_1(y_1) & \cdots & \phi_1(y_k) \\ \vdots & & \\ \phi_k(y_1) & \cdots & \phi_k(y_k) \end{vmatrix} \neq 0. \tag{2.11}$$

Thus, given arbitrary $b_1, \ldots, b_k$, there exists a unique $v \in V$ that satisfies $v(y_i) = b_i, i = 1, \ldots, k$.

THEOREM 2.7

Every extremal signature for a V which satisfies the Chebyshev condition is based on at least $k + 1$ points.

Proof. Suppose Σ based on $y_1, \ldots, y_r$ is an extremal signature for V and $r \leqslant k$. Then there exists $v \in V$ such that $v(y_1) = 1, v(y_i) = 0, i = 2, \ldots, r$. Hence, if $\zeta_1, \ldots, \zeta_r$ are weights associated with Σ, we have $\zeta_1 v(y_1) = 0$, which is impossible. Thus $r \geqslant k + 1$. ∎

In view of Exercise 2.2.6, then, every primitive extremal signature for a V that satisfies the Chebyshev condition is based on precisely $k + 1$ points in the real case and r points in the complex case, where $k + 1 \leqslant r \leqslant 2k + 1$. The

Chebyshev condition is intimately connected with the uniqueness of best approximations, as our next result shows.

THEOREM 2.8 (Haar)

Every $f \in C(B)$ has a unique best approximation out of V if, and only if, V satisfies the Chebyshev condition with respect to B.

Proof. (i) If $f \in V$ there is nothing to prove. Suppose that $f \notin V$. Let v_1 and

v_2 be best approximations to f out of V. Let $\Sigma\,[f - v_1]$, based on $y_1, \ldots, y_r$, be an extremal signature for V whose existence is affirmed by Theorem 2.6, with associated weights $\zeta_1, \ldots, \zeta_r$ satisfying

$$\sum_{i=1}^{r} |\zeta_i| = 1.$$

Then

$$\|f - v_1\| = \sum_{i=1}^{r} \zeta_i [f(y_i) - v_1(y_i)] = \sum_{i=1}^{r} \zeta_i [f(y_i) - v_2(y_i)]$$

$$\leqslant \sum_{i=1}^{r} |\zeta_i| \, |f(y_i) - v_2(y_i)| \leqslant \|f - v_2\|, \qquad (2.12)$$

and, since $\|f - v_2\| = \|f - v_1\|$, the inequalities in (2.12) are both equalities, which is possible only if $|f(y_i) - v_2(y_i)| = \|f - v_2\|, i = 1, \ldots, r$ and sgn $[f(y_i) - v_2(y_i)] = $ sgn $\zeta_i, i = 1, \ldots, r$. But then

$$f(y_i) - v_2(y_i) = f(y_i) - v_1(y_i), \qquad i = 1, \ldots, r$$

or $v_2(y_i) = v_1(y_i) = 0, i = 1, \ldots, r$, and $v_2 - v_1$ has $r \geqslant k + 1$ zeros in B. Since V satisfies the Chebyshev condition $v_2 - v_1 = 0$ or $v_2 = v_1$. The best approximation is unique.

(ii) Suppose that V does not satisfy the Chebyshev condition; i.e., there exists $v_0 \in V, v_0 \neq 0$, and distinct points $y_1, \ldots, y_k \in B$ such that $v_0(y_i) = 0$; $i = 1, \ldots, k$. Hence $\Delta = 0$ [cf. (2.11)] and

$$\sum_{j=1}^{k} a_j \phi_i(y_j) = 0, \qquad i = 1, \ldots, k,$$

has a nontrivial solution. Suppose that, after renumbering if necessary, $a_j \neq 0$, $j = 1, \ldots, r$, where $1 \leqslant r \leqslant k$; then

$$\sum_{j=1}^{r} a_j v(y_j) = 0, \qquad \text{all } v \in V. \qquad (2.13)$$

Suppose that $g \in C(B)$, $\|g\| = 1$, and $g(y_j) = $ sgn $\bar{a}_j, j = 1, \ldots, r$. Consider $f(y) = g(y)(1 - |\lambda v_0(y)|) + v_1(y)$, where v_1 is any element of V and $\|\lambda v_0\| < 1$. Then $f \in C(B)$, $\|f - v_1\| \leqslant 1$, and indeed $\|f - v_1\| = 1$ and $[f(y_j) - v_1(y_j)] = $ sgn $\bar{a}_j, j = 1, \ldots, r$. Put $\Sigma(y_j) = $ sgn $\bar{a}_j, j = 1, \ldots, r$. Then, in view of (2.13), Σ is an extremal signature for V, hence by Theorem 2.6 v_1 is a best approximation to f. Put $v_2 = v_1 - \lambda v_0$. Then $v_2 \in V$ and $f - v_2 = g(1 - |\lambda v_0|) + \lambda v_0$; hence $|f(y) - v_2(y)| \leqslant 1$ and $\|f - v_2\| = 1$. Thus v_2 is another best approximation to f. Uniqueness fails. ∎

Remark. Note that although the f we constructed is continuous, it is not differentiable.

2.4. Approximation on an Interval

We now further restrict our attention to the real case in which V satisfies the Chebyshev condition. In view of Theorem 2.7 and Exercises 2.2.7 and 2.2.9, we see that any distinct points $y_1, \ldots, y_{k+1}$ of B are the base of a unique extremal signature up to multiplication by -1. Moreover, if $B = [a, b]$, an interval of the real axis, the primitive extremal signatures are further restricted.

THEOREM 2.9

Suppose that $a \leqslant x_1 < x_2 < \cdots < x_{k+1} \leqslant b$; V satisfies the Chebyshev condition with respect to $[a, b]$ and consists of real-valued functions. Then the values $\Sigma(x_i)$, $i = 1, \ldots, k + 1$, of the "unique" extremal signature based on $x_1, \ldots, x_{k+1}$ alternate in sign.

Proof. It is enough to show that any weights $\zeta_1, \ldots, \zeta_{k+1}$ associated with Σ alternate in sign.

First we dispose of the case $k = 1$. If $k = 1$, $v \neq 0$ has no zero in $[x_1, x_2]$ and $\zeta_1 v(x_1) + \zeta_2 v(x_2) = 0$ then implies that $\zeta_1 \zeta_2 < 0$. Suppose that $k > 1$ and there exists j, $1 \leqslant j < k + 1$, such that $\zeta_j \zeta_{j+1} > 0$ and, say, $\zeta_j, \zeta_{j+1} > 0$. Then there exists q, $1 < q < k + 1$, such that either ζ_{q-1} or ζ_{q+1} is of the same sign as ζ_q, namely positive. (If $j > 1$, take $q = j$. If $j = 1$, take $q = 2$.) Now choose two positive numbers, α_{q-1} and α_{q+1}, such that $\alpha_{q-1}\zeta_{q-1} + \alpha_{q+1}\zeta_{q+1} > 0$, and let v_0 be the unique element of V that satisfies

$$v_0(x_{q-1}) = \alpha_{q-1}, \qquad v_0(x_{q+1}) = \alpha_{q+1},$$
$$v_0(x_i) = 0, \qquad i = 1, \ldots, k + 1, i \neq q - 1, q, q + 1.$$

Then by the definition of weights we have

$$\alpha_{q-1}\zeta_{q-1} + \zeta_q v_0(x_q) + \alpha_{q+1}\zeta_{q+1} = 0.$$

Thus $v_0(x_q) < 0$, while $v_0(x_{q+1}) > 0$; hence v_0 has two zeros in (x_{q-1}, x_{q+1}) for a total of at least k zeros in $[x_1, x_{k+1}]$. This contradicts the assumption that V satisfies the Chebyshev condition with respect to $[a, b]$. ∎

We next obtain, quite easily, the classical characterization theorem of best approximation on a closed subset of a real interval in the presence of the Chebyshev condition.

THEOREM 2.10

Let B denote the interval $[a, b]$ and suppose that V satisfies the Chebyshev *condition with respect to B. Let S be any closed subset of B (e.g., B, finitely many points of B). Then v_0 is the best approximation on S to $f \in C(S)$ ($f \notin V$) if, any only if, there exist $x_i \in E(f - v_0; S)$, $i = 1, \ldots, k + 1$, such that $x_1 < x_2 < \cdots < x_{k+1}$, and*

$$f(x_i) - v_0(x_i), \qquad i = 1, \ldots, k + 1,$$

alternate in sign.

Proof.   The Theorem follows immediately from Theorems 2.6 and 2.9.   ■

Remark.   Under the assumptions of Theorem 2.10 and in view of Corollary 2.6.1, the best approximation v_0 to f on S is also the best approximation to f on some set of $k + 1$ points of S. Moreover, if $Y: \{y_1, \ldots, y_{k+1}\}$ consists of distinct points of S and v_Y is the best approximation to f on Y, then $\|f - v_Y\|_Y \leqslant \|f - v_0\|$; that is to say, if we find the best approximation to f on a subset of $k + 1$ points of S which has the largest error (in norm) among all subsets of $k + 1$ points of S, that approximation is also best on all of S. This observation is the foundation of some frequently used algorithms for finding best approximations numerically. In the case that S is a finite set of points our search is reduced to examining the best approximations on the finite number of subsets of $k + 1$ points of S. This problem turns out to be tractable, as we shall see next.

Let X denote the distinct points $x_1, \ldots, x_{k+1}$ of $[a, b]$, numbered (for future convenience) this time as follows: $a \leqslant x_{k+1} < x_k < \cdots < x_1 \leqslant b$. Suppose that V satisfies the Chebyshev condition with respect to $[a, b]$. Consider the matrix

$$\Phi = \begin{pmatrix} \phi_1(x_1) & \cdots & \phi_k(x_1) \\ \vdots & & \\ \phi_1(x_{k+1}) & \cdots & \phi_k(x_{k+1}) \end{pmatrix}$$

and the determinant $\Delta_i (i = 1, \ldots, k + 1)$ of the square matrix obtained by deleting the ith row of Φ. No Δ_i is zero because of the Chebyshev condition and, indeed, they all have the same sign, for the determinant

$$\delta_i(t) = \begin{vmatrix} \phi_1(x_1) & \cdots & \phi_k(x_1) \\ \vdots & & \\ \phi_1(x_{i-1}) & \cdots & \phi_k(x_{i-1}) \\ \phi_1(t) & \cdots & \phi_k(t) \\ \phi_1(x_{i+2}) & \cdots & \phi_k(x_{i+2}) \\ \vdots & & \\ \phi_1(x_{k+1}) & \cdots & \phi_k(x_{k+1}) \end{vmatrix}$$

is a continuous function of t which is not zero for $x_{i+2} < t < x_{i-1}$; hence $\Delta_i = \delta_i(x_{i+1})$ and $\Delta_{i+1} = \delta_i(x_i)$ must have the same sign, and this holds for

$i = 1, \ldots, k + 1$ (with trivial modification of the argument for $i = 1, k + 1$). If we adjoin an arbitrary $(k + 1)$st column to Φ, the cofactors of the elements of this column are $(-1)^{k+i-1} \Delta_i, i = 1, \ldots, k + 1$. Moreover, these cofactors are orthogonal to each of the columns of Φ [because the dot product involved is merely the value of the determinant, with the particular column repeated as the $(k + 1)$st column, obtained by expanding by elements of the $(k + 1)$st column]; i.e.,

$$\sum_{i=1}^{k+1} (-1)^i \Delta_i \, \phi_j(x_i) = 0, \qquad j = 1, \ldots, k. \tag{2.14}$$

Equations 2.14 imply that

$$\Theta_i = \frac{(-1)^i \Delta_i}{\sum_{j=1}^{k+1} |\Delta_j|}, \qquad i = 1, \ldots, k + 1, \tag{2.15}$$

are weights associated with the extremal signature $\Sigma(x_i) = (-1)^i$, which satisfy

$$\sum_{i=1}^{k+1} |\Theta_i| = 1.$$

If v^* is the unique best approximation to $f \notin V$ on X, then in view of Theorem 2.10

$$\|f - v^*\|_X = \lambda \sum_{i=1}^{k+1} \Theta_i \, [f(x_i) - v^*(x_i)] = \lambda \sum_{i=1}^{k+1} \Theta_i \, f(x_i),$$

where

$$\lambda = \text{sgn} \sum_{i=1}^{k+1} \Theta_i \, f(x_i). \tag{2.16}$$

Thus we have proved the following:

THEOREM 2.11

For any f defined on X we have

$$\rho(f; X) = \rho = \|f - v^*\|_X = \left| \sum_{i=1}^{k+1} \Theta_i f(x_i) \right| \tag{2.17}$$

and v^* is determined by any k of the equations

$$v^*(x_i) = f(x_i) - \lambda \, \text{sgn} \, \Delta_i (-1)^i \rho, \qquad i = 1, \ldots, k + 1.$$

We now further specialize by taking V to be $\mathscr{P}_{k-1}$, real polynomials of degree at most $k - 1$, and B to be the interval I: $[-1, 1]$. (There is no loss in generality in restricting our attention to I, since results obtained are easily translated to any finite interval by means of a linear change of variables that leaves $\mathscr{P}_{k-1}$ invariant.) It is not easy to find explicitly the polynomial of best approximation to a given continuous $f(x)$. Of particular interest to us is the case $f(x) = x^k$, $(k \geqslant 1)$.

Let $p(x) \in \mathscr{P}_{k-1}$ be the best approximation to x^k on I out of $\mathscr{P}_{k-1}$. Then $r(x) = x^k - p(x) \in \mathscr{P}_k$, and $r(x)$ attains its maximum absolute value, $M > 0$, at $k + 1$ points satisfying $-1 \leqslant x_{k+1} < x_k < \cdots < x_1 \leqslant 1$, with alternating signs (though the latter fact is superfluous in determining p), according to Theorem 2.10. We claim that $x_1 = 1$ and $x_{k+1} = -1$, for otherwise $r(x)$ has a relative extremum at k interior points of I, which means that r' has k zeros, hence is identically zero, an impossibility [since the leading coefficient of $r(x)$ is 1]. Consider $M^2 - r^2(x) \in \mathscr{P}_{2k}$. $M^2 - r^2(x) \geqslant 0$ in I, hence has $x_2, \ldots, x_k$ as zeros of order, at least 2, and ± 1 as simple zeros, thus accounting for all its zeros; but $(1 - x^2)[r'(x)]^2 \in \mathscr{P}_{2k}$ has precisely the same zeros, hence is a constant multiple of $M^2 - r^2(x)$. The constant is determined by equating leading coefficients and we obtain the (Chebyshev) differential equation

$$(1 - x^2)(r'(x))^2 = k^2(M^2 - r^2(x)). \qquad (2.18)$$

We know that $r(-1) = \pm M$. Suppose that $r(-1) = -M$. Then we know that $r'(x) > 0$ in $[-1, x_k)$ and the differential equation becomes

$$\frac{r'(x)}{\sqrt{M^2 - r^2(x)}} = \frac{k}{\sqrt{1 - x^2}},$$

which has the solution arc cos $(r/M) = k$ arc cos $x + c = k\theta + c$, where $0 \leqslant \theta \leqslant \pi$, and cos $\theta = x$. Thus

$$r(x) = M \cos(k\theta + c).$$

Now $r(-1) = -M$ implies that $c = [(2j - 1) - k]\pi$ for some integer j, hence

$$r(x) = (-1)^{k+1}M \cos k\theta, \qquad 0 \leqslant \theta < \theta_k = \text{arc cos } x_k, \quad x = \cos \theta;$$

that is, $r(x) = (-1)^{k+1}M T_k(x)$ in $[-1, x_k)$. This is possible only if k is odd and $M = 2^{1-k}$, since the leading coefficient of r is 1. Also it must be that $x_k = \eta_{k-1}^{(k)}$, and it is clear that the solution

$$r(x) = \tilde{T}_k(x) \qquad (2.19)$$

can be continued in similar fashion to the whole interval I. An analogous examination of the case $r(-1) = M$ leads to the conclusion that k is even and that (2.1) remains valid. Thus the unique best approximation to x^k out of $\mathscr{P}_{k-1}$ on I is $x^k - \tilde{T}_k(x)$.

Note that we have arrived at another proof of Theorem 2.1. It has been a long voyage with some interest of its own, bringing us back to our starting point, but we observe that our second approach creates the Chebyshev polynomials out of the void while they are pulled out of a hat in Theorem 2.1.

In the course of the discussion just concluded we have established the following important characterization of the Chebyshev polynomials.

THEOREM 2.12

If $p \in \mathscr{P}_n$ and $|p|$ assumes its maximum on $[-1, 1]$, $\|p\|$, at $n + 1$ distinct points, then either p is a constant $(\pm \|p\|)$ or $p = \pm \|p\| T_n$, and the points are $\eta_j^{(n)}, j = 0, \ldots, n$.

Remark. If the interval $[-1, 1]$ is replaced by $[a, b]$, then $T_n(x)$ must be replaced by

$$T_n\left(\frac{2x - (a + b)}{b - a}\right)$$

in Theorem 2.12. In particular, when $a = 0$, $b = 1$, we have the Chebyshev polynomial for the interval $[0, 1]$, $T_n(2x - 1)$, which arises frequently enough in applications to warrant the special notation $T_n^*(x)$. Observe that in view of Exercise 1.1.6 $T_n^*(x^2) = T_{2n}(x)$.

EXERCISES 2.4

2.4.1. If $f(x)$ is an even (odd) function on $[-a, a]$, it has a best approximation out of $\mathscr{P}_n$ that is also even (odd).

Hint If $p(x)$ and $p(-x)$ are both best approximations, so is $[p(x) + p(-x)]/2$

2.4.2. What is the best approximation to x^{n+2} on $[-1, 1]$ out of $\mathscr{P}_n$?

2.4.3. If v^* is a best approximation to f out of V, then $v^* - v$ is a best approximation to $f - v$ out of V. Hence $E_V(f) = E_V(f - v)$, all $v \in V$.

2.4.4. If $f''(x) > 0$ on $[a, b]$, the best approximation to f out of $\mathscr{P}_1$ is

$$\frac{f(a)(b - c) + f(b)(c - a) + f(c)(b - a)}{2(b - a)} + \frac{f(b) - f(a)}{b - a}(x - c)$$

and

$$E_1(f) = \frac{f(a)(b - c) + f(b)(c - a) - f(c)(b - a)}{2(b - a)},$$

where c is the unique solution in $[a, b]$ of

$$f'(c) = \frac{f(b) - f(a)}{b - a}.$$

2.4.5. (Turán [1]) If

$$p(x) = x^n + a_{n-1}x^{n-1} + \cdots + a_0 = (x - z_1) \cdots (x - z_n),$$

then

$$\|p\| \geqslant 2^{-n+1} \prod_{|z_j|>1} |z_j|,$$

where an empty product is taken equal to 1.

Hint. Apply Corollary 2.1.1 to

$$p(x) \prod_{|z_j|>1} (x - z_j)^{-1}(1 - \bar{z}_j x).$$

2.4.6. According to (2.18), the differential equation

$$(1 - x^2)(y')^2 = n^2(1 - y^2) \tag{2.20}$$

has $y = \pm T_n(x)$ as solutions. Show that (2.20) has no other polynomial solution for $n > 0$.

Hint. If polynomial p satisfies (2.20), it satisfies $(1 - x^2)p'' - xp' + n^2 p = 0$ and $p(1) = \pm 1$. Now recall Exercise 1.5.4.

Problems 2.4.7–2.4.17 are set in the complex plane.

2.4.7. For any $\rho > 0$

$$\Sigma[\rho e^{2\pi ij/(k+1)}] = e^{-2\pi ij/(k+1)}, \qquad j = 0, 1, \ldots, k$$

is a primitive extremal signature for $\mathscr{P}_{k-1}$ (complex-valued polynomials).

Hint. Associated weights are $e^{2\pi ij/(k+1)}$.

2.4.8. Let C_ρ be the ellipse defined in the z-plane by

$$z = \tfrac{1}{2}\left(w + \frac{1}{w}\right), \qquad |w| = \rho, \qquad (0 < \rho \leqslant 1). \tag{2.21}$$

It has its foci at $(\pm 1, 0)$ and the sum of its major and minor axes is 2ρ. Let $w_j = \rho e^{j\pi i/k}$ and $z_j = (w_j + w_j^{-1})/2, j = 0, \ldots, 2k - 1$, be the corresponding points of C_ρ. Then

$$\Sigma(z_j) = (-1)^j, \qquad j = 0, \ldots, 2k - 1,$$

is an extremal signature for $\mathscr{P}_{k-1}$.

Hint. Associated weights are $(-1)^j$.

2.4.9. What happens when $\rho = 1$ in Exercise 2.4.8?

On each compact set B in the complex plane (consisting of $k + 1$ or more points) there exists a unique $p^* \in \mathscr{P}_k$ with leading coefficient one and minimum maximum modulus. We put $p^* = T_k(z; B)$ and call it the Chebyshev polynomial of degree k of B. We know that $T_k(x; I) = \tilde{T}_k(x)$.

2.4.10. If D denotes $|z| \leqslant 1$, $T_k(z; D) = z^k$.

2.4.11. $T_k(z; C_\rho) = \tilde{T}_k(z)$, $(0 < \rho \leqslant 1)$.

Hint. Use Exercise 2.4.8 and, recalling Exercise 1.1.1, note that $T_k(z) = (w^k + w^{-k})/2$, where z is given by (2.21).

2.4.12.

$$\max_{z \in C_\rho} |T_k(z; C_\rho)| = \frac{\rho^k + \rho^{-k}}{2^k}.$$

2.4.13. Show that all the zeros of $T_k(z; B)$ lie in $\hat{B}$, the convex hull of B.

If we put

$$\max_{z \in B} |T_k(z; B)| = m_k,$$

then, if B contains infinitely many points,

$$\delta(B) = \lim_{k \to \infty} m_k^{1/k}$$

exists and is called the *transfinite diameter* of B. Generalizations of the Chebyshev polynomial of B, based on the factorization $|T_k(z; B)| = |z - z_1| \ldots |z - z_k|$, have been made to other metric spaces (see Hille [1] and Friedman [1]).

2.4.14. $\delta(I) = \frac{1}{2}$; $\delta(D) = 1$; $\delta(C_\rho) = \dfrac{1}{2\rho}$.

2.4.15. If $0 < t < 1$, the function

$$w(z) = \frac{z - t}{1 - tz}$$

maps the circle $C: |z| = 1$ onto $|w| = 1$ so that, if $e^{i\phi}$ is the image of $e^{i\theta}$, $\phi(\theta)$ increases continuously and monotonely from 0 to 2π as θ increases from 0 to 2π.

2.4.16. Let $f(z) = 1/(1 - tz)$, $0 < t < 1$, and $p^*(z) = 1 + tz + \cdots + t^{n-1}z^{n-1} + (1 - t^2)^{-1} t^n z^n$; then $r(z) = f(z) - p^*(z) = (t^{n+1})(1 - t^2)^{-1} z^n (z - t)/(1 - tz)$ and $E(r; D) = C$.

2.4.17.   With the notation of Exercise 2.4.16, p^* is the best approximation to f on D out of $\mathscr{P}_n$ and $E_n(f) = t^{n+1}/(1 - t^2)$.

Hint.   Use the Kolmogorov criterion (Exercise 2.2.1). Note that $n\theta + \phi(\theta)$ increases from 0 to $(2n + 2)\pi$ as θ increases from 0 to 2π; hence there exist θ_j, $j = 0, 1, \ldots, 2n + 1$ such that $e^{i(n\theta_j + \phi(\theta_j))} = (-1)^j$. Consider only the points $e^{i\theta_j}$ of $E(r; D)$ in (2.9) and observe that Re $p(e^{i\theta})$ is a trigonometric polynomial of order n.

2.4.18.   If $0 < t < 1$, the best approximation to

$$g(x) = \frac{1 - tx}{1 + t^2 - 2tx}$$

on I: $[-1, 1]$ out of $\mathscr{P}_n$ is

$$p(x) = 1 + tT_1(x) + t^2 T_2(x) + \cdots + t^{n-1} T_{n-1}(x) + \frac{1}{1 - t^2} t^n T_n(x),$$

and $E_n(g; I) = t^{n+1}/(1 - t^2)$. The same result holds for $-1 < t < 1$.

Hint.   In Exercise 2.4.17 Re $r(e^{i\theta}) = [t^{n+1}/(1 - t^2)] \cos(n\theta + \phi)$, which assumes its maximum absolute value with alternating signs at $\theta_j, j = 0, \ldots, n + 1$.

2.4.19.   Find the best approximation out of $\mathscr{P}_n$ on I to $(x - \lambda)^{-1}$, where $\lambda > 1$.

2.4.20.   If $q \in \mathscr{P}_{n+1}$ and $\lambda > 1$, find the best approximation to $q(x)/(x - \lambda)$ out of $\mathscr{P}_n$ on $[-1, 1]$.

Hint.   Recall Exercise 2.4.3.

2.4.21.   If real V satisfies the Chebyshev condition with respect to $[a, b]$, there exists $v \in V$ satisfying $v(x) > 0, a \leqslant x \leqslant b$.

Hint.   If such a v does not exist, consider the best approximation to 1 out of V.

2.4.22.   Let X be the $k + 1$ points defined on p. 70 but suppose that V satisfies the Chebyshev condition only with respect to X. By appropriate modification of the discussion on p. 70 show that (2.17) still holds. How can v^* be determined now?

2.4.23.   The linear system

$$\sum_{j=1}^{k} a_{ij} t_j = f_i, \qquad i = 1, \ldots, k + 1$$

generally has no solution; it is overdetermined. However, we can ask for a best approximate solution, t^*, satisfying

$$\max_{1 \leqslant i \leqslant k+1} \left| \sum_{j=1}^{k} a_{ij} t_j^* - f_i \right| \leqslant \max_{1 \leqslant i \leqslant k+1} \left| \sum_{j=1}^{k} a_{ij} t_j - f_i \right|$$

for all t. If every $k \times k$ submatrix of $A = (a_{ij})$ is nonsingular, show that this problem is a special case of Exercise 2.4.22.

In problems 2.4.24–2.4.32 we are concerned with best approximation to given f on $X = \{x_1, \ldots, x_{k+1} / x_1 > x_2 > \cdots > x_{k+1}\}$ out of $V = \mathscr{P}_{k-1}$.

2.4.24. If $\omega(x) = (x - x_1) \cdots (x - x_{k+1})$, then (2.17) holds with

$$\Theta_i = \frac{(-1)^i / |\omega'(x_i)|}{\sum_{i=1}^{k+1} 1/|\omega'(x_i)|}, \qquad i = 1, \ldots, k+1. \tag{2.22}$$

2.4.25. Let

$$\tau = \frac{f(x_1, \ldots, x_{k+1})}{g(x_1, \ldots, x_{k+1})}$$

(recall the definition of divided differences in Exercise 1.3.7), where g is any function satisfying $g(x_i) = (-1)^i$, $i = 1, \ldots, k+1$. Show that $g(x_1, \ldots, x_{k+1}) < 0$, $\rho = |\tau|$ (ρ is defined in (2.17)), and if $|f(x_1, \ldots, x_{k+1})| \leq h(x_1, \ldots, x_{k+1})$ then $\rho(f; X) \leq \rho(h; X)$.

2.4.26. The best approximation to f on X out of $\mathscr{P}_{k-1}$ is given in terms of interpolating polynomials by

$$p^* = p - \tau q,$$

where $p = L_k(f, X)$ and $q = L_k(g, X)$, with τ and g as defined in Exercise 2.4.25.

2.4.27. The operator $\pi: f \to p^*$ is *linear* and satisfies $\pi^2 = \pi$.

2.4.28. Let $p_i \in \mathscr{P}_{k-1}, i = 1, \ldots, k+1$ satisfy $p_i(x_j) = f(x_j), j = 1, \ldots, k+1$, $j \neq i$. Then

$$p^* = \sum_{i=1}^{k+1} |\Theta_i| p_i,$$

where Θ_i is given by (2.22).

2.4.29. If $a = -1$ and $b = 1$

$$\rho(x^k; X) \leq \frac{1}{2^{k-1}}.$$

Thus

$$\sum_{j=1}^{k+1} \frac{1}{|\omega'(x_j)|} \geq 2^{k-1},$$

with equality in both cases if, and only if, $x_j = \eta_{j-1}^{(k)}, j = 1, \ldots, k+1$.

2.4.30.   If $a = -1$ and $b = 1$,

$$\rho(x^{k+1}; X) \leq \frac{1}{2^k},$$

and

$$\sum_{j=1}^{k+1} \frac{1}{|\omega'(x_j)|} \geq |x_1 + \cdots + x_{k+1}| 2^k.$$

Equality holds in both cases only if $x_j = \eta_{j-1}^{(k+1)}$, $j = 1, \ldots, k+1$, or $x_j = \eta_j^{(k+1)}$, $j = 1, \ldots, k+1$.

Hint.   If $f(x) = x^{k+1}$, then $f(x_1, \ldots, x_{k+1}) = x_1 + \cdots + x_{k+1}$.

2.4.31.   Show that if

$$U = \{\eta_0^{(k)}, \ldots, \eta_k^{(k)}\}$$

and

$$T = \{\xi_1^{(k+1)}, \ldots, \xi_{k+1}^{(k+1)}\}$$

we have

$$\rho(f; U) = \frac{1}{k} \left| \sum_{j=0}^{k} {}''(-1)^j f(\eta_j^{(k)}) \right| \tag{2.23}$$

and

$$\rho(f; T) = \sin \frac{\pi}{2(k+1)} \left| \sum_{j=1}^{k+1} (-1)^j \sin \frac{(2j-1)\pi}{2(k+1)} f(\xi_j^{(k+1)}) \right|.$$

2.4.32.   If $X \subset I: [-1, 1]$ and $f, h \in C^k(I)$, then

$|f(x_1, \ldots, x_{k+1})| \leq h(x_1, \ldots, x_{k+1})$ for every X if, and only if $|f^{(k)}(x)| \leq h^{(k)}(x)$, all $x \in I$.

Hint.   Recall Exercise 1.3.12.

2.4.33.   If $|f^{(k)}(x)| \leq h^{(k)}(x)$, all $x \in I$, then

$$E_{k-1}(f) \leq E_{k-1}(h)$$

on I.

Hint.   Exercises 2.4.32 and 2.4.25 plus the remark following Theorem 2.10.

2.4.34. On $[-1, 1]$

$$\frac{e^{-1}}{k!2^{k-1}} \leqslant E_{k-1}(e^x) \leqslant \frac{e}{k!2^{k-1}}.$$

(Compare the best approximation to the partial sum of the Taylor series!)

2.4.35. A trigonometric polynomial of degree n,

$$t(\theta) = \sum_{j=0}^{n} (a_j \cos j\theta + b_j \sin j\theta),$$

which assumes its maximum $\| t \|$ at $2n$ distinct values of $[0, 2\pi)$, is either a constant $(\pm \| t \|)$ or $\cos(n(\theta - \theta_0))$ for some θ_0.

Hint. Analog of Theorem 2.12.

Encouraged by our success in finding the polynomial with leading coefficient 1 that deviates least from zero on I (Theorem 2.1), let us consider fixing the next highest coefficient as well. Suppose $s \geqslant 0$ and p^* is the best approximation to $x^{k+1} - sx^k$ out of $\mathscr{P}_{k-1}$ on I. The polynomial

$$Z_{s,k}(x) = x^{k+1} - sx^k - p^*$$

is called a Zolotarev polynomial of order k and is a generalization of the Chebyshev polynomial (of order $k + 1$) to which it reduces when $s = 0$.

2.4.36. Show that

$$Z_{s,k}(x) = \frac{[1 + s/(k+1)]^{k+1}}{2^k} T_{k+1} \left[\frac{(k+1)x - s}{(k+1) + s} \right]$$

if

$$0 \leqslant s \leqslant (k+1) \tan^2 \frac{\pi}{2(k+1)}.$$

2.4.37. $Z_{-s,k}(x) = (-1)^{k+1} Z_{s,k}(-x)$.

2.4.38. Discuss $Z_{s,1}$ for all real s.

2.4.39. Discuss $Z_{s,2}$, for all real s.

Hint. Represent the parameter s in terms of a new parameter u by

$$s = \frac{3u}{2} - \frac{1}{2u}.$$

Theorem 2.1 suggests the problem of finding polynomials with leading coefficient 1 which deviate least from zero on I, where the deviation is measured in norms other than the uniform one. In the case of least-squares, with respect to the weight $(1 - x^2)^{-1/2}$, we saw in (1.109) that $\tilde{T}_n$ was once again the minimum deviator. More generally, as Szegö [1, p. 39] shows, for the least-squares norm with respect to a broad class of weight functions the appropriate orthogonal polynomial is the minimum deviator. The next set of exercises is concerned with the same problem in least first powers. If $|g|$ is integrable on I, we put

$$\|g\|_1 = \int_1^1 |g|\, dx.$$

2.4.40. If $p \in \mathscr{P}_{n-1}$ satisfies

$$\int_{-1}^1 \operatorname{sgn} [f(x) - p(x)]\, q(x)\, dx = 0 \tag{2.24}$$

for all $q \in \mathscr{P}_{n-1}$, then $\|f - p\|_1 \leqslant \|f - q\|_1$ for all $q \in \mathscr{P}_{n-1}$.

2.4.41. Show that

$$\int_{-\pi}^\pi \operatorname{sgn} [\sin (n + 1)\theta]\, \sin k\theta\, d\theta = 0, \qquad k = 1, \ldots, n.$$

2.4.42. If $r(x) = x^n + c_{n-1}x^{n-1} + \cdots + c_0$, then

$$2^{1-n} = \|\tilde{U}_n\|_1 \leqslant \|r\|_1. \tag{2.25}$$

Hint. $\tilde{U}_n$ is the Chebyshev polynomial of the second kind [see (1.23)] normalized so that its leading coefficient is 1. Use Exercises 2.4.40 and 2.4.41 and note that every $q \in \mathscr{P}_{n-1}$ can be written $q = a_0 U_0 + \cdots + a_{n-1}U_{n-1}$.

2.4.43. If $p \in \mathscr{P}_{n-1}$ satisfies (2.24) and is thus a least-first-power approximation to f, then

$$\|f - p\|_1 = \int_{-1}^1 f \operatorname{sgn} [f - p]\, dx.$$

2.4.44. Equality holds in (2.25) only if $r = \tilde{U}_n$.

Hint. If $r = x^n - p_1$ and equality holds in (2.25), then

$$\int_{-1}^1 r(x) \operatorname{sgn} U_n(x)\, dx = \int_{-1}^1 |r(x)|\, dx,$$

which implies that $r(x)$ changes sign precisely as $U_n(x)$ does at $\eta_1^{(n+1)}, \ldots, \eta_n^{(n+1)}$.

2.4.45. If $f - p$ changes sign only at $\eta_1^{(n+1)}, \ldots, \eta_n^{(n+1)}$ and $p \in \mathscr{P}_{n-1}$, then p is a least-first-power approximation to f on I.

2.4.46. If $w(x)$ is a weight function and $t \geq 1$, then if $p \in \mathscr{P}_{n-1}$ satisfies

$$\int_{-1}^{1} \operatorname{sgn}(f-p)\,|f-p|^{t-1}q(x)\,w(x)\,dx = 0$$

for all $q \in \mathscr{P}_{n-1}$,

$$\int_{-1}^{1} |f-p|^t w\,dx \leq \int_{-1}^{1} |f-q|^t w\,dx$$

for all $q \in \mathscr{P}_{n-1}$.

2.4.47. If $q(x) = x^n + a_{n-1}x^{n-1} + \cdots + a_0$, then

$$\int_{-1}^{1} |\tilde{T}_n(x)|^t\,\frac{dx}{\sqrt{1-x^2}} \leq \int_{-1}^{1} |q(x)|^t\,\frac{dx}{\sqrt{1-x^2}}$$

for all $t \geq 1$.

Another generalization of Theorem 2.1 is to minimize

$$\max_{-1 \leq x \leq 1} s(x)\,|x^n + a_{n-1}x^{n-1} + \cdots + a_0\,|, \tag{2.26}$$

where $s(x)$ is a positive continuous function on I. (Our general theory informs us that $s(x)x^n$ has a unique best approximation out of the space spanned by $s(x)$, $xs(x), \ldots, x^{n-1}s(x)$ on I.) Such problems have been solved for special choices of s. We put

$$v_k(x) = \prod_{j=1}^{k}\left(1 - \frac{x}{c_j}\right), \qquad k \geq 1;\ v_0 = 1,$$

where the c_j are any points of the complex plane such that $v_k(x)$ is positive on I. Now consider the case that $s = v_k^{-1}$. By means of the mapping (2.21)

$$c_j = \frac{1}{2}\left(w_j + \frac{1}{w_j}\right), \qquad j = 1, \ldots, k, \tag{2.27}$$

where $|w_j| < 1$. Suppose that $q_k(w) = (w-w_1)\cdots(w-w_k)$.

2.4.48. If $w = e^{i\phi}, 0 \leq \phi \leq \pi$, then if $x = (w+w^{-1})/2$

$$q_k(w)\,q_k(w^{-1}) = \left[\prod_{j=1}^{k}(1+w_j^2)\right]v_k(x) = W_k v_k(x) \tag{2.28}$$

Hint. Equation 2.27 implies $w_j^2 - 2c_j w_j + 1 = 0$ and the product on the left is therefore a polynomial in x whose zeros are $c_1, \ldots, c_k$.

2.4.49. If $n > k$, $w = e^{i\phi}$, $0 \leqslant \phi \leqslant \pi$ and $x = (w + w^{-1})/2$,

$$T_n(x; v_k) = \frac{W_k}{2^n} \left[w^{n-2k} \frac{q_k(w)}{q_k(w^{-1})} + w^{2k-n} \frac{q_k(w^{-1})}{q_k(w)} \right] v_k(x)$$

$$= x^n + b_{n-1} x^{n-1} + \cdots + b_0.$$

Hint. Use Exercise 2.4.48 and $T_j(x) = (w^j + w^{-j})/2$.

2.4.50. When $s = v_k^{-1}$, the unique minimum of (2.26) is assumed for $a_j = b_j$ (defined in Exercise 2.4.49), $j = 0, \ldots, n-1$; i.e., if $p \in \mathscr{P}_{n-1}$ is the best approximation to x^n with respect to weight $[v_k(x)]^{-1}$, then $x^n - p(x) = T_n(x; v_k)$.

B. MAXIMIZING LINEAR FUNCTIONALS ON $\mathscr{P}_n$.

2.5. An Interpolation Formula for Linear Functionals

We shall continue to use the notation of Section 2.2 throughout this section; however, let B be a compact set in real m-space and V a k-dimensional subspace of the *real-valued* continuous functions on B, $C(B)$.

DEFINITION 2.9

A (real) linear functional on V is a function, F, with domain V and range in the real numbers which satisfies $F(au + bv) = aF(u) + bF(v)$ for every $u, v, \in V$ and any real numbers a, b.

Examples of linear functionals are $Fv = v(y_0)$, where y_0 is a given point of B (point evaluation functional) or

$$Fv = \int_B v.$$

If $\phi_1, \ldots, \phi_k$ is a basis for V, then a linear functional F is completely specified by its values at $\phi_1, \ldots, \phi_k$, for, if $F\phi_j = c_j, j = 1, \ldots, k$, and $v = a_1\phi_1 + \cdots + a_k\phi_k$, then $Fv = a_1 c_1 + \cdots + a_k c_k$. Furthermore, the set of a such that $v = a_1\phi_1 + \cdots + a_k\phi_k$ satisfies $\|v\| \leqslant 1$ is compact; thus $|Fv|$ is a continuous function on $\|v\| \leqslant 1$ and assumes its maximum there.

DEFINITION 2.10

If F is a linear functional on V and

$$\max_{\|v\| \leqslant 1} |Fv| = M,$$

M is called the norm of F, written $\|F\|$, and there exists v^* satisfying $\|v^*\| = 1$ such that $Fv^* = \|F\|$. Such v^* are called extremal elements, or extremal, for F.

When V is $\mathscr{P}_n$, Theorem 2.1 says that if

$$Fv = \frac{v^{(n)}(0)}{n!}$$

then $\|F\| = 2^{n-1}$ and T_n is the only extremal element for F. Our goal now is to examine a large class of linear functionals on $\mathscr{P}_n$ for which the Chebyshev polynomials are extremal elements. Our main tool in this program is an "interpolation formula" for linear functionals on V which is itself another consequence of Theorem 2.4. Before stating the formula we need a little more information about linear functionals.

The set of $v \in V$ such that $Fv = 0$ is called the null-space of F. The null-space of F is a $k - 1$-dimensional subspace of V (see Exercise 2.5.2).

THEOREM 2.13

Let $F\ (\neq 0)$ be a real linear functional on V. Then there exist points $y_1, \ldots, y_r$ of B and nonzero real numbers $\alpha_1, \ldots, \alpha_r$, with $r \leqslant k$, such that for every $v \in V$

$$Fv = \sum_{j=1}^{r} \alpha_j v(y_j) \tag{2.29a}$$

and

$$\|F\| = \sum_{j=1}^{r} |\alpha_j|. \tag{2.29b}$$

Proof. Let v^* be an extremal element for F. If $k = 1$, then any $v \in V$ can be written cv^*; hence $Fv = cFv^* = c\|F\|$. Let y_1 be a point of B such that $v^*(y_1) = \epsilon$, where $\epsilon = \pm 1$; then $v(y_1) = c\epsilon$ and $c = \epsilon v(y_1)$. Thus $Fv = \epsilon\|F\|v(y_1)$ and the Theorem is proved by choosing $\alpha_1 = \epsilon\|F\|$.

Suppose, then, that $k > 1$. Let V_0 denote the null space of F, which has dimension $k - 1$ and thus contains nonzero elements. If $v_0 \in V_0$, then, since $|F(v^* + v_0)| \leqslant \|v^* + v_0\|\,\|F\|$ and $F(v^* + v_0) = \|F\|$, we have $\|v^* + v_0\| \geqslant 1 = \|v^*\|$. Invoking Theorem 2.4, with V_0 playing the role of V and g replaced by v^*, we obtain the existence of $r \leqslant k$ points $y_1, \ldots, y_r$ of $E(v^*; B)$ and positive numbers $\lambda_1, \ldots, \lambda_r$ such that

$$\sum_{i=1}^{r} \lambda_i v^*(y_i)\, v_0(y_i) = 0, \tag{2.30}$$

all $v_0 \in V_0$.

Suppose that $v \in V$; then $v_0 = (Fv)v^* - (Fv^*)v \in V_0$, and by substituting this v_0 in (2.30) we obtain

$$(Fv) \sum_{i=1}^{r} \lambda_i [v^*(y_i)]^2 = Fv^* \sum_{i=1}^{r} \lambda_i v^*(y_i) \, v(y_i)$$

or, since $|v^*(y_i)| = \|v^*\| = 1$ and $Fv^* = \|F\|$,

$$(Fv) \sum_{i=1}^{r} \lambda_i = \|F\| \sum_{i=1}^{r} [\lambda_i \, \text{sgn} \, v^*(y_i)] \, v(y_i).$$

The theorem is now proved by putting

$$\alpha_i = \frac{\lambda_i \, \text{sgn} \, v^*(y_i)}{\sum_{i=1}^{r} \lambda_i} \|F\|. \quad \blacksquare$$

A representation of F of the form (2.29) we call *canonical.* An important observation for us is the following:

THEOREM 2.14

If v^* is extremal for F, then for any canonical representation (2.29)

$$v^*(y_j) = \text{sgn} \, \alpha_j, \qquad j = 1, \ldots, r. \tag{2.31}$$

Proof.

$$\sum_{j=1}^{r} |\alpha_j| = \|F\| = Fv^* = \sum_{j=1}^{r} \alpha_j \, v^*(y_j),$$

and the theorem follows, since $\|v^*\| = 1$. $\blacksquare$

EXERCISES 2.5

2.5.1. Show that the set of linear functionals on V is itself a k-dimensional normed linear space, the norm being that given in Definition 2.10.

2.5.2. If $F_1, \ldots, F_s (s < k)$ are linearly independent linear functionals on V and N_i is the null space of F_i, $i = 1, \ldots, s$, then

$$\bigcap_{i=1}^{s} N_i$$

is a $k - s$-dimensional subspace of V.

2.5.3. If $F, F_1, \ldots, F_s$ are linear functionals on V with respective null spaces $N, N_1, \ldots, N_s$ and

$$\bigcap_{i=1}^{s} N_i \subset N,$$

then F is a linear combination of $F_1, \ldots, F_s$.

2.5.4. Show that no y_i in (2.29a) is a common zero of all elements of V.

2.5.5. If $1 \in V$, then 1 is extremal for F if, and only if, F has a canonical representation with $\alpha_j > 0, j = 1, \ldots, r$.

2.5.6. If $1 \in V$, then 1 is extremal for F if and only if F is a *positive* linear functional (i.e., if $v(y) \geq 0$, all $y \in B$, then $Fv \geq 0$).

2.5.7. Let $x = (x_1, \ldots, x_m)$ denote a point in real m space and let $V = P(m, n)$ be the space of polynomials in x of degree at most n, i.e., all $x_1^{j_1} \cdots x_m^{j_m}$ with nonnegative integers $j_1, \ldots, j_m$ satisfying $j_1 + \cdots + j_m \leq n$ form a basis for

$P(m, n)$. It is not hard to see that the dimension of $P(m, n)$ is $k = \binom{m + n}{m}$. (The

dimension is the number of ways of putting n balls in $m + 1$ bins. To obtain this number we need only calculate the number of ways of choosing m "partitions" (=interior bin walls) among $n + m$ objects arranged on a line). Show that if B is a compact set in m space there is a numerical integration formula

$$\int_B f \, dx = \sum_{i=1}^{r} a_i f(x^{(i)}), \qquad r \leq k, \tag{2.32}$$

with $x^{(i)} \in B$ and $a_i > 0, i = 1, \ldots, r$, valid for $f \in P(m, n)$.

Hint. $Ff = \int f$ is a positive linear functional.

2.5.8. Let x denote a point in real m space, let B be a compact set in m space, and V a k-dimensional subset of $C(B)$ spanned by $\phi_1(x), \ldots, \phi_k(x)$ with $\phi_1(x) > 0$ throughout B. Then, if F is a positive linear functional on V,

$$Fv = \sum_{i=1}^{r} a_i v(x^{(i)}), \qquad r \leq k,$$

with $x^{(i)} \in B$ and $a_i > 0, i = 1, \ldots, r$, holds for all $v \in V$. (This result, which implies the result of Exercise 2.5.7, is due to Tchakaloff [1].)

Hint. $Gf = F(\phi_1 f)$ is a positive linear functional on the span of 1, $\phi_2/\phi_1, \ldots, \phi_k/\phi_1$.

2.5.9. If F is a *strictly positive* linear functional on $\mathscr{P}_n$ (i.e., if $p \geq 0, p \neq 0$, then $Fp > 0$) then

$$\left[\frac{n}{2}\right] < r \leq n + 1$$

in any canonical representation of F.

2.5.10. Put

$$m = \left[\frac{n}{2}\right] + 1.$$

Let $x_1, \ldots, x_m$ be the distinct zeros of the Legendre polynomial $P_m(x)$ (cf. p. 40) in $(-1, 1)$. Show that there is a numerical integration formula (Gaussian quadrature formula)

$$\int_{-1}^{1} f(x)\, dx = \sum_{i=1}^{m} a_i f(x_i)$$

with $a_i > 0, i = 1, \ldots, m$, valid for $f \in \mathscr{P}_n$.

Hint. The integral in question is a strictly positive linear functional on $\mathscr{P}_n$. If $p \in \mathscr{P}_n$ is zero at each x_i, then $\int p\, dx = 0$ in view of the orthogonality of the Legendre polynomials. Now apply Exercise 2.5.3.

2.5.11. Let $V(F) = \{v \in V / Fv = 1\}$. Show that (i) if v^* is extremal for F then $\bar{v} = v^*/\|F\|$ satisfies

$$\rho_F(V) = \min_{v \in V(F)} \|v\| = \|\bar{v}\|. \tag{2.33}$$

(ii) If $\bar{v} \in V(F)$ satisfies (2.33), then $\bar{v}/\|\bar{v}\|$ is extremal for F. Thus $\|F\|\rho_F(V) = 1$; v^* is unique if, and only if, $\bar{v}$ is unique.

2.6. Linear Functionals on $\mathscr{P}_n$.

Henceforth we take $V = \mathscr{P}_n$. We can now say something about the uniqueness of a canonical representation of F.

THEOREM 2.15

If ± 1, is not the unique extremal for F, then F has a unique canonical representation.

Proof. Suppose that F has two canonical representations

$$Fp = \sum_{j=1}^{r} \alpha_j\, p(x_j) = \sum_{j=1}^{s} \beta_j\, p(y_j), \qquad p \in \mathscr{P}_n \tag{2.34}$$

with

$$\sum_{j=1}^{r} |\alpha_j| = \sum_{j=1}^{s} |\beta_j| = \|F\|$$

and say $r \geqslant s$.

Let $p^* \neq \pm 1$ be extremal for F; then, in view of Theorem 2.14, $|p^*(x_j)| = |p^*(y_j)| = 1, j = 1, \ldots, r; i = 1, \ldots, s$. Hence the set $\{x_1, \ldots, x_r, y_1, \ldots, y_s\}$ contains $k \leqslant n + 1$ distinct points, $z_1, \ldots, z_k$.

We claim that $\{x_1, \ldots, x_r\} = \{y_1, \ldots, y_s\}$. Suppose not, say, $x_l \neq y_i$, $i = 1, \ldots, s$. Let $z_k = x_l$. Then

$$q(x) = \prod_{i=1}^{k-1} (x - z_i) \in \mathscr{P}_n,$$

$q(x_l) \neq 0$, and $q(y_i) = 0$, $i = 1, \ldots, s$. Thus $Fq = \alpha_l \, q(x_l) = 0$, according to (2.34), a contradiction. Hence $r = s$ and, after renumbering if necessary, $x_j = y_i$, $j = 1, \ldots, r$. Now consider

$$p_j(x) = \prod_{\substack{i=1 \\ i \neq j}}^{r} (x - x_i), \qquad j = 1, \ldots, r,$$

$p_j(x) \in \mathscr{P}_n$, and $p_j(x_i) \neq 0$; $Fp_j = \alpha_j \, p_j(x_j) = \beta_j \, p_j(x_j)$ implies that $\alpha_j = \beta_j$, $j = 1, \ldots, r$. ∎

Remark 1. The requirement that neither 1 nor -1 be a unique extremal for F is essential for a canonical representation to be unique, as the following example shows. Consider $\mathscr{P}_2$ and the functional $F(a_0 + a_1 x + a_2 x^2) = 3a_0 + 2a_2$. Then

$$Fp = p(-1) + p(0) + p(1) = \frac{3}{2} p\left(-\frac{\sqrt{6}}{3}\right) + \frac{3}{2} p\left(\frac{\sqrt{6}}{3}\right),$$

are both canonical representations, and clearly 1 is the unique extremal for F.

Remark 2. If 1 is an extremal for F, then F has a canonical representation containing a preassigned point, t, of B, if, and only if,

$$\min_{p \in \mathscr{Q}_n} \Gamma p - b > 0,$$

where $\mathscr{Q}_n = \{p \in \mathscr{P}_n / p(t) = 1 \text{ and } p(x) \geqslant 0, x \in B\}$. To see this we apply Tchakaloff's theorem (Exercise 2.5.8) to the linear functional $Fq - b \, q(t)$, which is positive in view of the definition of b and the positivity of F (Exercise 2.5.6), and obtain

$$Fq = b \, q(t) + \sum_{i=1}^{r} a_i \, q(x_i), \qquad r \leqslant n + 1,$$

with $a_i > 0$, $i = 1, \ldots, r$. If $Fp^* = b$, $p^* \in \mathscr{Q}_n$,

$$b = Fp^* = b + \sum_{i=1}^{r} a_i \, p^*(x_i),$$

and therefore $p^*(x_i) = 0$, $i = 1, \ldots, r$, which implies that $r \leqslant n$ (indeed, since

$p^* \geqslant 0$, $2r \leqslant n$), and we have the required canonical representation. Conversely, if $b = 0$ and

$$Fq = cq(t) + \sum_{i=1}^{r} c_i\, q(x_i), \qquad r \leqslant n,$$

with c, c_i positive and $x_i \neq t$, putting $q = p^*$ (where p^* is defined above), yields a contradiction.

A result of which we shall make major use provides sufficient conditions for the Chebyshev polynomial to be an extremal of a linear functional. Henceforth we take $B = [-1, 1]$.

THEOREM 2.16

If $r = n + 1$ in a canonical representation of F, then the extremals for F are either $\pm T_n$ or ± 1. In the former case we may take $y_j = \eta_{j-1}^{(n)}$, $j = 1, \ldots, n + 1$, and $\alpha_1, \ldots, \alpha_{n+1}$ alternate in sign.

Proof.   Immediate consequence of Theorems 2.14 and 2.12.   ∎

Remark.   However, $r = n + 1$ is not a necessary condition for T_n to be an extremal, for, consider

$$Fp = \sum_{j \in J} (-1)^j p(\eta_j^{(n)}),$$

where J is some subset of $\{0, 1, \ldots, n\}$. Clearly, if $\|p\| = 1$, then $|Fp| \leqslant |J|^{\dagger}$, whereas

$$FT_n = \sum_{j \in J} 1 = |J|.$$

Hence T_n can be an extremal of a linear functional with $1 \leqslant r \leqslant n + 1$ in its canonical representation.

Next, we turn to the question of the uniqueness of extremals for a given linear functional. We define a function on I, $e(x)$ by

$$e(x) = \begin{cases} 2 & -1 < x < 1, \\ 1 & x = \pm 1. \end{cases}$$

---

† If S is a finite set, $|S|$ denotes its cardinality.

THEOREM 2.17

If

$$\sum_{j=1}^{r} e(y_j) > n \qquad (2.35)$$

for some canonical representation of F, then F has a unique extremal.

Proof. If p, q are both extremals for F, then sgn $\alpha_j = p(y_j) = q(y_j)$ so that $p(y_j) - q(y_j) = 0, j = 1, \ldots, r$ and $p'(y_j) - q'(y_j) = 0$, if $-1 < y_j < 1$. Thus $p - q$ has zeros of total multiplicity greater than n, in view of (2.35), and $p = q$. ∎

Results of a converse nature are not so neat. However, we have the following:

THEOREM 2.18

If 1 is extremal for F and

$$\sum_{j=1}^{r} e(y_j) \leqslant n$$

for some canonical representation, then 1 is not the unique extremal for F.

Proof. Suppose that $y_1 < y_2 < \cdots < y_r \leqslant 1$. If $y_r = 1$, set

$$q(x) = (1 - x) \prod_{j=1}^{r-1} (x - y_j)^{e(y_j)},$$

while if $y_r < 1$ take

$$q(x) = \prod_{j=1}^{r} (x - y_j)^{e(y_j)}.$$

In either case $q \in \mathscr{P}_n$ and $q(x) \geqslant 0$ in I. Choose $A > 0$ and so small that $\|Aq\| < 1$, then $p = 1 - Aq \in \mathscr{P}_n$ and $\|p\| = 1$, but $p(y_j) = 1, j = 1, \ldots, r$; hence

$$Fp = \sum_{j=1}^{r} \alpha_j = \|F\|.$$

Since $q \neq 0, p \neq 1$. ∎

Remark. The same result holds if -1 is extremal for F. If neither ± 1 is extremal for F, then (2.35) is not necessary for uniqueness. Consider the remark following Theorem 2.16 with $n = 4$ and $J = \{2, 3\}$. Clearly, T_4 is extremal and ± 1 is not.

Suppose that $p \in \mathscr{P}_4$ is also extremal so that $\| p \| = 1, p(\eta_2) = 1$, and $p(\eta_3) = -1$. Then $q = p - T_4$ satisfies $q(\eta_j) = 0, q'(\eta_j) = 0, j = 2, 3$, hence

$$q = c(x - \eta_2)^2 (x - \eta_3)^2,$$

but $|p(1)| = |1 - q| \leq 1$ requires $c \geq 0$, whereas $|p(\eta_1)| = |-1 - q| \leq 1$ requires $c \leq 0$. Hence $c = 0$ and $p = T_4$.

To obtain a necessary condition for uniqueness we proceed as follows. If $p \in \mathscr{P}_n$, let $N(p)$ denote the total multiplicity with which the values 1 and -1 are assumed by p in I. (Thus $N(T_n) = 2n$.)

THEOREM 2.19

If F has a unique extremal, p, then $N(p) > n$.

Proof. We show that if $N(p) \leq n$ then F has an extremal other than p. Suppose that $N(p) \leq n$.

Let $x_1, \ldots, x_s$ satisfying $-1 \leq x_1 < \cdots < x_s \leq 1$ be all the points of I at which $p(x) = 1$ and let m_i be the multiplicity of the zero of $1 - p(x)$ at x_i. Let $z_1, \ldots, z_t$ satisfying $-1 \leq z_1 < \cdots < z_t \leq 1$ be all the points of I at which $p(x) = -1$ and let n_i be the multiplicity of the zero of $1 + p(x)$ at z_i. (We need only consider the case $s \geq 1, t \geq 1$, for if $p(x)$ omits the value -1, say, on I, then either $p = 1$ and nonuniqueness follows from Theorem 2.18 or $p \neq 1$ and 1 is another extremal for F. Similarly, if p omits the value 1.)

Put

$$q_1(x) = c_1 \prod_{i=1}^{s} (x - x_i)^{m_i}$$

and

$$q_2(x) = c_2 \prod_{i=1}^{t} (x - z_i)^{n_i},$$

where c_1 and c_2 are chosen so that there is some subinterval of I in which $q_1(x) > 0$ and another in which $q_2(x) > 0$. Thus $1 - p(x) = q_1(x) p_1(x)$ where $p_1(x) > 0$ throughout I and $1 + p(x) = q_2(x) p_2(x)$ where $p_2(x) > 0$ throughout I. Note, also, that, since $0 \leq 1 - p(x)$ and $0 \leq 1 + p(x), q_1(x) \geq 0$ and $q_2(x) > 0$ for all $x \in I$, whereas $q = q_1 q_2 \in \mathscr{P}_n$ in view of our assumption that $N(p) \leq n$. Let

$$b = \min_{x \in I} \frac{p_1(x)}{q_2(x)};$$

then $b > 0$, and if $0 < a \leq b$ we assert that $p(x) + aq(x) = r(x)$ satisfies $\| r \| = 1$, for $a \leqslant b$ implies that $aq(x) \leqslant q_1(x) p_1(x)$, hence $r(x) \leqslant 1$ for all $x \in I$, whereas $a > 0$ implies $aq(x) \geqslant -q_2(x) p_2(x)$ and $r(x) \geqslant -1$. Since $r(x_i) = p(x_i)$ and $r(z_i) = p(z_i)$, r is another extremal. ∎

EXERCISES 2.6

2.6.1. The remark following Theorem 2.16 is not convincing unless we know that the F defined there does *not* have another canonical representation involving $n + 1$ points. Show that this is the case.

2.6.2. If $Fx^i = 0$ for all odd i, $0 \leqslant i \leqslant n$, show that F has an even extremal. If $Fx^i = 0$ for all even i, $0 \leqslant i \leqslant n$, show that F has an odd extremal.

2.6.3. If $Fx^i = 0$, either for all odd i or for all even i, $0 \leqslant i \leqslant n$, and neither of ± 1 is an extremal for F, show that F has a canonical representation in which the points are symmetric about zero. In the second case zero is not one of the points. Also, the coefficients [the α_j in (2.29a)] of symmetric points are equal.

Hint. Put $x^2 = t$ and consider either polynomials of degree at most $[n/2]$ or linear combinations of $t^{1/2}, t^{3/2}, \ldots, t^{[(n+1)/2]-1/2}$ on $[0, 1]$.

2.6.4. If the null space of F satisfies the Chebyshev condition, the only extremals of F are either $\pm T_n$ or ± 1.

2.6.5. Consider the linear functional on $\mathscr{P}_2$, $Fp = p(-1) - p(0)$. Show that $\|F\| = 2$ and T_2 is the unique extremal for F.

2.6.6. Show that the null space V_0 of the functional F defined in Exercise 2.6.5 does not satisfy the Chebyshev condition.

2.6.7. Let V_0 be as in Exercise 2.6.6. Show that there is a *unique* best approximation to x^2 out of V_0 on $[-1, 1]$.

The set of $p \in \mathscr{P}_n$ satisfying $\|p\| \leqslant 1$ ($\|\cdot\|$ is the maximum norm on I) is a compact convex set, B_n, the unit ball in $\mathscr{P}_n$. A point $p \in B_n$ is an *extreme point* of B_n if $p = (p_1 + p_2)/2$ with $p_1, p_2 \in B_n$ implies $p = p_1 = p_2$. Let us denote the set of extreme points of B_n by $E(B_n)$.

2.6.8. $1 \in E(B_n)$ for $n = 0, 1, 2, \ldots$.

2.6.9. If $p \in B_n$ and $p(x_0) = 1$, where $-1 < x_0 < 1$, then there exists a positive integer m such that $p(x) = 1 - (x - x_0)^{2m} r(x)$, where $r(x) \geqslant 0$ in I and $r(x_0) > 0$.

2.6.10. If $p = (p_1 + p_2)/2$, where $p_1, p_2 \in B_n$, $p(x_0) = \epsilon$, and $p^{(k)}(x_0) = 0$, $k = 1, \ldots, k_0 - 1$, $p^{(k_0)}(x_0) \neq 0$, where $x_0 \in I$ and $\epsilon = \pm 1$. Then, for $j = 1, 2$, $p_j(x_0) = \epsilon$ and $p_j^{(k)}(x_0) = 0$, $k = 1, 2, \ldots, k_0 - 1$.

2.6.11. If $p \in B_n$ then $p \in E(B_n)$ if, and only if, $N(p) > n$. (Notation as in Theorem 2.19.)

Hint. (i) The "if" part follows from Exercise 2.6.9, 10; (ii) "only if" by construction following that given in the proof of Theorem 2.19.

2.6.12. $T_n \in E(B_k), n \leqslant k \leqslant 2n - 1$ but $T_n \notin E(B_{2n})$.

2.6.13.   Suppose 1 is an extremal for F (hence F is a positive linear functional); then the following are equivalent:

(i) $\Sigma_{i=1}^r e(y_i) \leqslant n$ for some canonical representation of F.
(ii) 1 is not the unique extremal for F.
(iii) F is not strictly positive (see Exercise 2.5.9).
(iv) F has a unique canonical representation.

Hint.   Remark 2 following Theorem 2.15 can be used to show that (iv) implies (iii).

2.7.   Some Examples in Which the Chebyshev Polynomials Are Extremal

We want next to use Theorem 2.16 in order to conclude that the Chebyshev polynomial is the extremal for certain specific functionals. Indeed, we shall examine a series of functionals that exemplify the following general scheme. Let C_n denote the (convex) subset of $\mathscr{P}_n$ consisting of p satisfying

$$\max_{j=0,\ldots,n} |p(\eta_j^{(n)})| \leqslant 1.$$

THEOREM 2.20

Let F be a linear functional on $\mathscr{P}_n$ such that (i) *neither ± 1 is an extremal for F,* (ii) $p \in \mathscr{P}_n, p \neq 0,$ *having n distinct zeros in I, implies $Fp \neq 0$. Then*

$$|Fp| \leqslant |FT_n|, \qquad p \in C_n, \tag{2.36}$$

with equality holding if, and only if, $p = \pm T_n$.

Proof.   Let

$$Fp = \sum_{j=1}^r \alpha_j \, p(y_j)$$

be a canonical representation of F. If $r \leqslant n$, there exists $p_0 \in \mathscr{P}_n, p_0 \neq 0$, satisfying $p_0(y_j) = 0, j = 1, \ldots, r$; hence $Fp_0 = 0$, contradicting the hypothesis (if $r < n$, choose p_0 to have zeros at any $n - r$ points of I other than the y_j). Thus $r = n + 1$ and, in view of Theorem 2.16, $y_j = \eta_{j-1}^{(n)}, j = 1, \ldots, n + 1$, and $\|F\| = |FT_n|$, but if $p \in C_n$

$$|Fp| \leqslant \sum_{j=1}^{n+1} |\alpha_j| \, |p(\eta_{j-1}^{(n)})| \leqslant \sum_{j=1}^{n+1} |\alpha_j| = \|F\| = |FT_n|. \qquad \blacksquare$$

Remark.   The theorem remains true, of course, with the condition, $p \in C_n$

strengthened to $p \in B_n$ (i.e., $\| p \| \leq 1$). It is in this weaker form that results of the kind we are about to give occur most frequently in the literature. In some of the examples we rely directly on Theorem 2.13 with V other than $\mathscr{P}_n$, but the line of argument is analogous to an application of Theorem 2.20.

1. *Growth Outside the Interval.* If $p \in C_n$, then

$$|p^{(j)}(t)| \leq |T_n^{(j)}(t)| \tag{2.37}$$

for $|t| \geq 1$ and $j = 0, 1, \ldots, n$. Equality is possible in (2.37) for $j \geq 1$ or $|t| > 1$ only if $p = \pm T_n$ (cf. Exercise 1.5.11).

To establish (2.37) we put $Fp = p^{(j)}(t)$. If $j = 0$ and $t = \pm 1$, (2.37) is trivially true (with equality holding for all $p \in C_n$ satisfying $|p(t)| = 1$). Suppose that $j \geq 1$ or $|t| > 1$; then neither of ± 1 is extremal for F, since $|F1| < |FT_n|$. Also, if $p \neq 0$ has n distinct zeros in I, it has no zero in $|t| > 1$ and, by Rolle's theorem, $p^{(j)}$ has no zero in $|t| \geq 1$; that is, $Fp \neq 0$. (i) and (ii) in the hypotheses of Theorem 2.20 are thus in force and our result follows from Theorem 2.20.

Remark. In particular, taking $j = n$, we see that if $p \in C_n$ has leading coefficient a_n then $|a_n| \leq 2^{n-1}$, with equality only for $p = \pm T_n$. One consequence of this observation is the following generalization of Theorem 2.1 ($n > 0$). If

$$p(x) = x^n + a_{n-1}x^{n-1} + \cdots + a_0,$$

then

$$\max_{j=0,\ldots,n} |p(\eta_j^{(n)})| \geq \max_{j=0,\ldots,n} |\tilde{T}_n(\eta_j^{(n)})| = 2^{1-n},$$

with equality only for $p = \pm\tilde{T}$, for if $p + +\tilde{T}_n$ and

$$\max_{j=0,\ldots,n} |p(\eta_j^{(n)})| \leq 2^{1-n}$$

then

$$2^{n-1}p \in C_n.$$

Another consequence of the same observation is an interesting characterization of the Chebyshev polynomials due to DeVore [1].

If $p(x) = c(x - x_1) \cdots (x - x_n)$ has all its zeros in $[-1, 1]$ and satisfies

$$|p(\eta_j^{(n)})| = 1, \qquad j = 0, \ldots, n$$

then $p = \pm T_n$. To see this note that

$$1 = |p(\eta_0)\, p^2(\eta_1) \cdots p^2(\eta_{n-1})\, p(\eta_n)| = |p(\eta_0) \cdots p(\eta_n)|\,|p(\eta_1) \cdots$$
$$p(\eta_{n-1})|$$

$$= \left| c \right|^{n+1} \prod_{j=1}^{n} \left| \prod_{i=0}^{n} (\eta_i - x_j) \right| \cdot |c|^{n-1} \prod_{j=1}^{n} \left| \prod_{i=1}^{n-1} (\eta_i - x_j) \right|$$

$$= \frac{|c|^{2n}}{[n^2 2^{2(n-1)}]^n} \prod_{j=1}^{n} (1 - x_j^2)[T_n'(x_j)]^2.$$

But in view of the Chebyshev differential equation (2.18)

$$(1 - x_j^2)[T_n'(x_j)]^2 = n^2(1 - T_n^2(x_j)) \leqslant n^2;$$

hence

$$1 \leqslant \left(\frac{|c|}{2^{n-1}} \right)^{2n}$$

or

$$|c| \geqslant 2^{n-1}.$$

Since $p \in C_n$, we know that $|c| \leqslant 2^{n-1}$; hence $|c| = 2^{n-1}$ and $p = \pm T_n$.

2.   *Size of Coefficients.* If $p = a_0 + \cdots + a_n x^n$ and $p \in C_n$, then

$$\left| \sum_{j=1}^{[n/2]} a_{n-2j} \right| \leqslant \left| \sum_{j=1}^{[n/2]} t_{n-2j}^{(n)} \right|, \qquad i = 0, 1, \ldots, \left[\frac{n}{2} \right]. \tag{2.38}$$

Let us put

$$F_{i,n} p = \sum_{j=i}^{[n/2]} a_{n-2j}, \qquad i = 0, \ldots, \left[\frac{n}{2} \right].$$

Our aim is to show that $\pm T_n$ is extremal for $F_{i,n}$. If $i = 0$ or $[n/2]$, (2.38) holds trivially with equality for any $p \in C_n$ satisfying $p(1) = \pm 1$ or $p(0) = \pm 1$, respectively. This disposes of the cases $n = 1, 2$ and we need only establish (2.38) if $n \geqslant 3$ and $0 < i < [n/2]$.

Suppose that $a_n = 1, p \in \mathscr{P}_n$ is even for even n or odd for odd n, and either (i) p has n distinct zeros in I or (ii) p has a zero of order 2 at zero and $n - 2$ other distinct zeros in I. Then $F_{0,n} p = p(1) \geqslant 0$, since $a_n = 1$, and we claim that if $n \geqslant 4$ is even

$$(-1)^{n-i} F_{i,n} p > 0, \qquad i = 1, \ldots, \frac{n}{2} - 1, \tag{2.39a}$$

and

$$(-1)^{n/2} F_{n/2,n} \, p \geqslant 0, \tag{2.39b}$$

with equality if, and only if, $p(0) = 0$, whereas if $n \geqslant 3$ is odd

$$(-1)^{n-i} F_{i,n} p < 0, \qquad i = 1, \ldots, \frac{n-1}{2}. \tag{2.40}$$

We verify this claim by mathematical induction on n. Consider first even n. Suppose that $n = 4$. Let c be the smallest nonnegative zero of p. Then $p = (x^2 - c^2)(x^2 - a^2)$, where $0 < |a| \leqslant 1$, $F_{2,4} p = a^2 c^2 \geqslant 0$, with equality if, and only if, $c = 0$, and $F_{1,4} p = a^2 c^2 - (a^2 + c^2) = c^2(a^2 - 1) - a^2 < 0$ verifying (2.39a,b). Suppose that (2.39a,b) hold for $n \geqslant 4$ and suppose even $p \in \mathscr{P}_{n+2}$ has $a_{n+2} = 1$ and satisfies (i) or (ii) above. Let c be the smallest nonnegative zero of p. Then $p = (x^2 - c^2)r$, where $r \in \mathscr{P}_n$ is even, has leading coefficient 1, satisfies (i), and

$$(-1)^{n+2-i} F_{i,n+2} p = (-1)^{n-i} F_{i,n} r + c^2 (-1)^{n-(i-1)} F_{i-1,n} r$$

so that (2.39a,b) hold and (2.39a,b) are therefore valid for all even n. In the case that $n(\geqslant 3)$ is odd a similar argument establishes (2.40).

Now let a canonical representation of $F_{i,n} (n \geqslant 3, i = 1, \ldots, [(n-1)/2])$ with points symmetric about the origin be

$$F_{i,n} q = \sum_{j=1}^{r} \alpha_j \, q(y_j), \qquad q \in \mathscr{P}_n, \tag{2.41}$$

the existence of such a canonical representation being assured by Exercise 2.6.3, provided that ± 1 is not an extremal of $F_{i,n}$; but they are clearly not extremals for odd n, and for even n

$$F_{i,n} 1 = 1 < | F_{i,n} T_{n-2(i+1)} | = 2^{n-2i+3} - 1.$$

If $r \leqslant n$, there exists $p \in \mathscr{P}_n$, even for even n or odd for odd n, satisfying (i) or (ii) and taking the value zero at $y_1, \ldots, y_r$ such that, according to (2.39) or (2.40), $F_{i,n} p \neq 0$. Thus $r = n + 1$ and either $\pm T_n$ or ± 1 is extremal for $F_{i,n}$; but we have just seen that ± 1 is not an extremal for $F_{i,n}$, hence $y_j = \eta_{j-1}^{(n)}, j = 1, \ldots, n + 1$ and (2.38) is established, with equality only for $p = \pm T_n$, if $n \geqslant 3$ and $0 < i < [n/2]$.

Remark 1. If $p = a_0 + a_1 x + \cdots + a_n x^n, p \in B_n$, and we put

$$A_i(x) = \sum_{j=i}^{[n/2]} a_{n-2j} x^{n-2j}$$

for each $i = 0, \ldots, [n/2]$. Then, if, for some i,

$$\max_{x \in I} |A_i(x)| = |A_i(x_0)|,$$

the polynomial $q = p(x_0 x) \in C_n$ and so, in view of (2.38) applied to q,

$$\|A_i(x)\| = \left| \sum_{j=i}^{[n/2]} a_{n-2j} x_0^{n-2j} \right| \leqslant \left| \sum_{j=i}^{[n/2]} t_{n-2j}^{(n)} \right|, \qquad i = 0, \ldots, [n/2]. \tag{2.42}$$

In particular, we have, taking $a_{n-2j} = t_{n-2j}^{(n)}$

$$|t_0^{(n)} + \cdots + t_i^{(n)}| = \|t_0^{(n)} + \cdots + t_i^{(n)} x^i\|, \qquad i = 0, \ldots, n.$$

In a similar vein, if we know that

$$|a_i + \cdots + a_n| \leqslant |b_i + \cdots + b_n|$$

for all $p = a_0 + \cdots + a_n x^n \in B_n$, *which are even for even n or odd for odd n,* then $\|a_i x^i + \cdots + a_n x^n\| \leqslant |b_i + \cdots + b_n|$. Moreover, we note that

$$a_i + \cdots + a_n = p(1) - (a_0 + \cdots + a_{i-1}),$$

so that $|a_i + \cdots + a_n| \leqslant 1 + |t_0^{(n)} + \cdots + t_{i-1}^{(n)}|$; but we observed in Exercise 1.2.19 that the numbers $t_0^{(n)} + \cdots + t_{n-2j}^{(n)}, j = 0, \ldots, [n/2]$ alternate in sign. Therefore, since $t_0^{(n)} + \cdots + t_{i-1}^{(n)} = 1 - (t_i^{(n)} + \cdots + t_n^{(n)})$ is negative for $[n/2]$ of the indices $i = 1, \ldots, n+1$, and when that is the case $1 + |t_0^{(n)} + \cdots + t_{i-1}^{(n)}| = |t_i^{(n)} + \cdots + t_n^{(n)}|$, we conclude that at least "half" the time

$$|a_i + \cdots + a_n| \leqslant |t_i^{(n)} + \cdots + t_n^{(n)}|, \tag{2.43}$$

hence $\|a_i x^i + \cdots + a_n x^n\| \leqslant |t_i^{(n)} + \cdots + t_n^{(n)}|$. Indeed, Reimer [1] describes the relatively few cases in which (2.43) fails to hold.

Note that (2.42) is no longer true if we require only $p \in C_n$. Consider, for example, n an even integer greater than 2 and

$$p(x) = 1 - \prod_{j=0}^{(n/2-1)} ([\eta_j^{(n)}]^2 - x^2);$$

$p \in C_n$, but, since $p(\eta_i) = 1$ and $p'(\eta_i) \neq 0$ for some $j \neq 0$, $\|p\| > 1$ and $p \notin B_n$. Also $\|p\| = \|A_{n/2}(x)\| > |t_0^{(n)} + \cdots + t_n^{(n)}|$.

Remark 2. Suppose $p = a_0 + a_2 x + \cdots + a_n x^n$. If $n - j$ is even (or zero) and $p \in C_n$, then

$$|a_j| \leqslant |t_j^{(n)}|. \tag{2.44}$$

If $n - j$ is odd and $p \in C_{n-1}$, then

$$|a_j| \leqslant |t_j^{(n-1)}|. \tag{2.45}$$

Equality in (2.44) occurs only if $p = \pm T_n$ for $j > 0$ and equality in (2.45) occurs only if $p = \pm T_{n-1}$ for $j > 0$ and $n > 2$.

Proof. If $j = 0$ (2.44) and (2.45) are trivially true, with equality for any p satisfying $p(0) = \pm 1$. Suppose $j > 0$. Let $F_j p = a_j$. We treat several cases.

(a) $n - j$ even (or zero).

(1) n even.

$$a_j = F_j p = F_{i,n}p - F_{i+1,n}p, \qquad i = \frac{n-j}{2},$$

hence

$$(-1)^{n-i}F_j p = (-1)^n {}^{-i}F_{i,n}p + (-1)^{n-(i+1)}F_{i+1,n}p. \tag{2.46}$$

Since $F_j x^k = 0$ for all odd k, F_j has an even extremal according to Exercise 2.6.2. With F_j restricted to even polynomials (2.39a,b) together with (2.46) imply that

$$(-1)^{(n+j)/2}F_j p > 0, \qquad j = 2, 4, \ldots, n,$$

provided that $a_n = 1$ and either (i) or (ii) (p. 94) holds. Thus we conclude, as before, that $\pm T_n$ is a unique extremal, as ± 1 cannot be extremals, and (2.44) is established.

(2) n odd. An analogous treatment based on (2.40) establishes (2.44).

(b) $n - j$ odd.

F_j has an even extremal, if j is even, an odd extremal if j is odd. In either case it has an extremal in $\mathscr{P}_{n-1}$, but F_j restricted to $\mathscr{P}_n$ takes us back to (a), since $n - 1 - j$ is even (or zero), and $\pm T_{n-1}$ are the only possible extremals of F_j in $\mathscr{P}_{n-1}$. Moreover, in view of Exercise 2.6.3 F_j has a symmetric canonical representation, namely,

$$F_j p = \sum_{j=1}^{n} \alpha_j p(\eta_j^{(n-1)}), \tag{2.47}$$

and it is unique for $p \in \mathscr{P}_{n-1}$. Equation 2.47 is also canonical for $p \in \mathscr{P}_n$, for, since x^n is odd for odd n and even for even n, the right-hand side of (2.47) is zero when $p = x^n$. In view of Theorems 2.17 and 2.19 $\pm T_{n-1}$ is the unique extremal if $n > 2$, whereas for $n = 2$ one of $\pm T_{n-1}$ is an extremal but there are others. ∎

3. *The Tau Method.* Let $Q_n = \{p \in \mathscr{P}_n / p(0) = 1\}$. We wish to consider the problem of minimizing

$$\| p - p' \| \tag{2.48}$$

for all $p \in Q_n$. If we put $p - p' = q$, with $q(x) = b_0 + \cdots + b_n x^n$ and $p(x) = a_0 + \cdots + a_n x^n$, then

$$a_j = \frac{1}{j!} \sum_{i=j}^{n} i! \, b_i, \qquad j = 0, \ldots, n, \tag{2.49}$$

and the condition $p(0) = 1$, i.e., $a_0 = 1$, is equivalent to

$$Gq = \sum_{i=0}^{n} i! \, b_i = 1. \tag{2.50}$$

Thus our problem is to minimize $\| q \|$ among all $q \in \mathscr{P}_n$ that satisfy (2.50). In view of Exercise 2.5.11 it suffices to find an extremal for the linear functional G, defined in (2.50). Suppose that $n > 1$. (If $n = 0$, the original problem is trivial. If $n = 1$, $p(x) = 1 + \lambda x$, hence $q(x) = 1 + \lambda x - \lambda$ so that $q(1) = 1$ and $\| q \| \geqslant 1$, but, if $0 \leqslant \lambda \leqslant 1$, $\| q \| = 1$. Thus any $p(x) = 1 + \lambda x, 0 \leqslant \lambda \leqslant 1$, solves our problem.) We wish to apply Theorem 2.20 to G. Neither of ± 1 is extremal for G, since $Gx^n = n! > 1$. We claim that (ii) in Theorem 2.20 also holds. Indeed, we show more. If (for $n \geqslant 1$) $p \in \mathscr{P}_n$ has leading coefficient 1, n real zeros $x_1 \leqslant x_2 \leqslant \cdots \leqslant x_n \leqslant 1$, and $p \neq (x - 1)^n$, then, if $p(x) = c_0 + c_1 x + \cdots + c_n x^n$,

$$G_j p = \sum_{i=j}^{n} i! \, c_i > 0, \qquad j = 0, \ldots, n - 1. \tag{2.51}$$

We establish (2.51) be mathematical induction on n. If $n = 1$, $p = c_0 + x$ and $-c_0 < 1$, so that $G_0 p = c_0 + 1 > 0$, and (2.51) is proved. Suppose that (2.51) holds for n and $r_{n+1} \in \mathscr{P}_{n+1}$ satisfies

$$r_{n+1}(x) = (x - a) \, r_n(x),$$

where $a \leqslant 1$ and r_n satisfies the inductive hypothesis. Let $B_i = r_n^{(i)}(0)$ and $B_i' = r_{n+1}^{(i)}(0)$; then

$$B_i' = -aB_i + iB_{i-1}, \qquad i = 0, \ldots, n + 1,$$

and if $0 \leqslant j \leqslant n$

$$G_j r_{n+1} = \sum_{i=j}^{n+1} B_i' = \sum_{i=j+1}^{n} S_i + (1 - a)S_j + jS_{j-1},$$

where

$$S_k = \sum_{i=k}^{n} B_i, \qquad (S_{n+1} = 0).$$

By the inductive hypothesis $S_k > 0, k = 0, \ldots, n - 1$, and $B_n > 0$, since the leading coefficient of r_n is 1; hence $G_j p_{n+1} > 0$ and all is shown.

According to Theorem 2.20, therefore, the unique extremal for G is $\pm T_n$. If we put

$$\frac{1}{\tau_n} = GT_n = \sum_{i=0}^{n} T_n^{(i)}(0), \tag{2.52}$$

then, since T_n has a positive leading coefficient, $GT_n = G_0 T_n > 0$, and the unique extremal for G is T_n. Now Exercise 2.5.11 reveals that the unique solution of our minimum problem is

$$q_n(x) = \tau_n T_n(x),$$

or the unique $p_n \in Q_n$ that minimizes (2.48) is obtained from

$$p_n(x) - p_n'(x) = \tau_n T_n(x). \tag{2.53}$$

Also

$$\min_{p \in Q_n} \| p - p' \| = \| p_n - p_n' \| = \tau_n.$$

Note that in view of (2.51) and (2.49) all the coefficients of p_n are positive. If we put

$$s_k(x) = \sum_{j=0}^{k} \frac{x^j}{j!},$$

i.e., s_k is the kth partial sum of e^x, then $s_k(x) - s_k'(x) = x^k/k!$ and therefore

$$p_n(x) = \frac{\sum_{k=0}^{n} T_n^{(k)}(0) s_k(x)}{\sum_{k=0}^{n} T_n^{(k)}(0)} = \frac{\sum_{k=0}^{n} k! t_k^{(n)} s_k(x)}{\sum_{k=0}^{n} k! t_k^{(n)}}. \tag{2.54}$$

To see this observe that the polynomial defined in (2.54) is in Q_n and satisfies (2.53). Thus the solution of our problem turns out to be an "average" of the partial sums of e^x.

$y = e^x$ is the unique solution of $y - y' = 0, y(0) = 1$, hence minimizing (2.48) seems a reasonable way of obtaining a polynomial approximation to e^x on I. Indeed, the idea of choosing p to satisfy (2.53) and then choosing τ_n so that $p(0) = 1$ exemplifies the *tau method* of *Lanczos* [1].

Let us go a little further and see how good $p_n(x)$ is as an approximation of e^x. Consider $p'_{n+1} = u \in \mathscr{P}_n$. We shall show that u is "nearly" a best approximation out of $\mathscr{P}_n$ to e^x, not only on I but in the closed elliptical domain D_ρ consisting of the points of the ellipse C_ρ (see Exercise 2.4.8) and the points inside C_ρ, $0 < \rho \leqslant 1$. The reason we approximate by $u = p'_{n+1}$ rather than p_n is that p'_{n+1} is close to p_{n+1}, and so we get the advantage of a polynomial of degree $n + 1$ with a polynomial of degree n.

If $z \in C_\rho$, we solve the linear differential equation (2.53) subject to $p_{n+1}(0) = 1$ and obtain

$$e^z - p_{n+1}(z) = \tau_{n+1} \, e^z \int_0^z e^{-t} T_{n+1}(t) \, dt, \tag{2.55}$$

the path of integration being the line segment joining 0 and z. Differentiating (2.55) yields

$$e^z - u(z) = \tau_{n+1}\left(T_{n+1}(z) + e^z \int_0^z e^{-t} T_{n+1}(t) \, dt \right). \tag{2.56}$$

Now if $z \in C_\rho$, then

$$|z| \leqslant \frac{\rho + \rho^{-1}}{2} = \lambda \tag{2.57}$$

and (cf. Exercise 2.4.11)

$$|T_{n+1}(z)| \leqslant \frac{\rho^{n+1} + \rho^{-(n+1)}}{2}.$$

Also, since $\rho^t + \rho^{-t}$ is a concave increasing function of t for $t > 0$, it is not hard to see that, recalling Exercise 1.1.4,

$$W_{n+1}(z) = \int_0^z T_{n+1}(t) \, dt = \frac{1}{2}\left[\frac{T_{n+2}(z)}{n+2} - \frac{T_n(z)}{n} \right] + \cos\frac{n\pi}{2} \frac{n+1}{n(n+2)}$$

satisfies

$$|W_{n+1}(z)| \leqslant \frac{2}{n} \frac{\rho^{n+1} + \rho^{-(n+1)}}{2}.$$

Integrating by parts now yields

$$\left| e^z \int_0^z e^{-t} T_{n+1}(t) \, dt \right| \leqslant \frac{\beta}{n} \frac{\rho^{n+1} + \rho^{-(n+1)}}{2}, \tag{2.58}$$

where

$$\beta = 2(1 + \lambda e^{2\lambda}),$$

and λ is defined in (2.57). Thus for every $z \in D_\rho$

$$|e^z - u(z)| \leqslant \tau_{n+1}\left(1 + \frac{\beta}{n}\right)\frac{\rho^{n+1} + \rho^{-(n+1)}}{2}, \tag{2.59}$$

according to the maximum modulus principle.

In Exercise 2.4.8, with $k = n + 1$, an extremal signature based on points of C_ρ, $z_0, \ldots, z_{2n+1}$, is described. Putting $z = z_i$ in (2.56) yields

$$e^{z_i} - u(z_i) = \tau_{n+1}\left[(-1)^i \frac{\rho^{n+1} + \rho^{-(n+1)}}{2} + e^{z_i}\int_0^{z_i} e^{-t}T_{n+1}(t)\,dt\right],$$

and therefore

$$\mathrm{Re}\,(e^{z_i} - u(z_i))(-1)^i = \tau_{n+1}\left[\frac{\rho^{n+1} + \rho^{-(n+1)}}{2}\right.$$

$$\left. + (-1)^i\,\mathrm{Re}\,e^{z_i}\int_0^{z_i} e^{-t}T_{n+1}(t)\,dt\right],$$

but for each $i = 0, \ldots, 2n + 1$,

$$\mathrm{Re}\,e^{z_i}\int_0^{z_i} e^{-t}T_{n+1}(t)\,dt \leqslant \left|e^{z_i}\int_0^{z_i} e^{-t}T_{n+1}(t)\,dt\right| \leqslant \frac{\beta}{n}\frac{\rho^{n+1} + \rho^{-(n+1)}}{2},$$

in view of (2.58), hence

$$\min_i \mathrm{Re}\,(e^{z_i} - u(z_i))(-1)^i \geqslant \tau_{n+1}\left(1 - \frac{\beta}{n}\right)\frac{\rho^{n+1} + \rho^{-(n+1)}}{2}.$$

Corollary 2.6.2 now implies that

$$E_n(e^x; D_\rho) = E_n(e^x; C_\rho) \geqslant \tau_{n+1}\left(1 - \frac{\beta}{n}\right)\frac{\rho^{n+1} + \rho^{-(n+1)}}{2}. \tag{2.60}$$

Formulas 2.59 and 2.60 reveal that $u(z)$ is arbitrarily close to a best approximation for n sufficiently large.

Since

$$\frac{1}{\tau_n} = \sum_{j=0}^{\lceil n/2 \rceil} (n - 2j)!\, t_{n-2j}^{(n)}$$

and the explicit formula for the t_{n-2j} [see (1.96)] yields

$$(n - 2j)! \, t_{n-2j} + (n - 2(j + 1))! \, t_{n-2(j+1)} \begin{cases} > 0, & j = 0, 2, 4, \ldots, \\ < 0, & j = 1, 3, 5, \ldots, \end{cases}$$

we obtain

$$\frac{1}{\tau_n} > n! \, t_n + (n - 2)! \, t_{n-2} = n! \, 2^{n-3} \left(4 - \frac{1}{n-1} \right)$$

and

$$\frac{1}{\tau_n} < n! \, t_n = n! \, 2^{n-1}.$$

Thus

$$\frac{1}{(n + 1)! \, 2^n} < \tau_{n+1} < \frac{1}{(n + 1)! \, 2^n (1 - 4/n)}. \tag{2.61}$$

If the inequality (2.61) is used in (2.59) and (2.60) and we consider the interval I, i.e., $\rho = 1$, we obtain

$$\frac{1}{2^n (n + 1)!} \left[1 - \frac{2(1 + e^2)}{n} \right] < E_n(e^x; I) < \frac{1}{2^n (n + 1)!}$$

$$\times \left[1 + \frac{2(1 + e^2)}{n} \right] \left(\frac{1}{1 - 4/n} \right),$$

which, for n large enough, is significantly better than Exercise 2.4.34.

4. *Size of the Derivative.* Suppose that

$$t(\theta) = \sum_{j=0}^{n} (a_j \cos j\theta + b_j \sin j\theta),$$

a trigonometric polynomial of degree n, satisfies

$$| t(\theta_j) | \leqslant 1, \qquad j = 1, \ldots, 2n, \tag{2.62}$$

where $\theta_j = (2j - 1)\pi/2n, j = 1, \ldots, 2n$, then

$$| t'(0) | \leqslant n, \tag{2.63}$$

with equality only if $t = \pm \sin n\theta$.

To see this we apply Theorem 2.13 with $V = \mathcal{T}_n$, the $2n + 1$-dimensional space consisting of all trigonometric polynomials of degree, at most, n and

$Ft = t'(0)$. Let a canonical representation of F be

$$t'(0) = \sum_{j=1}^{r} \alpha_j t(y_j) \tag{2.64}$$

$(0 \leqslant y_j < 2\pi, r \leqslant 2n + 1)$. If $r < 2n$, then, since there exists (nonzero) $t_0 \in \mathcal{T}_n$ vanishing at $2n$ distinct points of $[0, 2\pi)$, including zero, we have $t_0'(0) = 0$ and t_0 has a total of $2n + 1$ zeros in $[0, 2\pi]$, a contradiction. Therefore $r \geqslant 2n$. Thus (see Exercise 2.4.35) the only possible extremals of F are ± 1, $\cos n(\theta - \theta_0)$. But ± 1 are not extremals, since their derivative is zero (hence, incidentally, $r = 2n$, for, if $r = 2n + 1$, ± 1 are the only possible extremals since for no nonconstant $t \in \mathcal{T}_n$ does $|t|$ assume its maximum at $2n + 1$ distinct points of $[0, 2\pi)$.). If $t = \cos n(\theta - \theta_0)$, then $t'(0) = n \sin n\theta_0$; hence the only possible extremals are $t = \cos n(\theta - [(2i - 1)\pi/2n]) = \pm \sin n\theta$. Thus in (2.64) $r = 2n$, $y_j = (2j - 1)\pi/2n = \theta_j$, and (2.63) follows at once.

Next, suppose we fix $\phi, 0 \leqslant \phi \leqslant 2\pi$ and put

$$t_\phi(\theta) = t(\theta + \phi).$$

Clearly, $t_\phi \in \mathcal{T}_n$, and if we assume that

$$|t(\theta_j + \phi)| \leqslant 1, \qquad j = 1, \ldots, 2n, \tag{2.65}$$

then, according to (2.63), we have $|t_\phi'(0)| \leqslant n$, i.e.,

$$|t'(\phi)| \leqslant n, \tag{2.66}$$

with equality only if $t(\theta) = \pm \sin n(\theta - \phi)$. Thus we have shown in particular that if $t \in \mathcal{T}_n$ and

$$\|t\| = \max_{0 \leqslant \theta \leqslant 2\pi} |t(\theta)| \leqslant 1 \tag{2.67}$$

then

$$\|t'\| \leqslant n, \tag{2.68}$$

with equality only for polynomials of the form $t(\theta) = \sin n(\theta - \phi)$. [The hypothesis (2.67) cannot be replaced by the less stringent (2.62), as the example $t(\theta) = \sin n\theta + 2 \cos n\theta$ shows.]

Remark 1. If $p(x) = a_0 + \cdots + a_n x^n$, then $t(\theta) = p(\cos \theta) \in \mathcal{T}_n$. Thus, if $p \in B_n$, then $\|t\| \leqslant 1$ and

$$|t'(\theta)| = |p'(x) \sin \theta| = |p'(x)(1 - x^2)^{1/2}| \leqslant n$$

or

$$|p'(x)| \leqslant \frac{n}{(1 - x^2)^{1/2}}, \qquad -1 < x < 1. \tag{2.69}$$

If $t \in \mathcal{T}_n$ and

$$\left| t\left(\frac{i\pi}{n}\right) \right| \leqslant 1, \qquad i = 0, \ldots, 2n - 1,$$

$t(\theta + \theta_k)$ satisfies (2.62) and we conclude that

$$| t'(\theta_k) | \leqslant n, \qquad k = 1, \ldots, 2n,$$

with equality for any k only if $t = \pm \cos n\theta$. Thus, if $p \in C_n$,

$$| p'(\xi_i^{(n)}) | \leqslant \frac{n}{\sqrt{1 - (\xi_i^{(n)})^2}} = | T_n'(\xi_i^{(n)}) |, \qquad i = 1, \ldots, n, \tag{2.70}$$

and equality is possible for any i only for $p = \pm T_n$. Thus we have obtained another proof of Exercise 1.5.8.

Also

$$| p^{(k)}(u_i) | \leqslant | T_n^{(k)}(u_i) |, \tag{2.71}$$

for $1 \leqslant k < n$, where $u_1, \ldots, u_{n-(k-1)}$ are the zeros of $T_n^{(k-1)}$, with equality possible for any i only if $p = \pm T_n$. (This is Exercise 1.5.9.) We establish this by mathematical induction on k. The case $k = 1$ is just (2.70). Suppose that (2.71) holds. The Lagrange interpolation formula gives

$$p^{(k)}(x) = \sum_{i=1}^{n-(k-1)} \frac{p^{(k)}(u_i)}{T_n^{(k)}(u_i)} \frac{T_n^{(k-1)}(x)}{x - u_i};$$

hence, if v is any zero of $T_n^{(k)}$,

$$p^{(k+1)}(v) = -T_n^{(k-1)}(v) \sum_{i=1}^{n-(k-1)} \frac{p^{(k)}(u_i)}{T_n^{(k)}(u_i)} \frac{1}{(v - u_i)^2}, \tag{2.72}$$

and

$$| p^{(k+1)}(v) | \leqslant | T_n^{(k-1)}(v) | \sum_{i=1}^{n-(k-1)} \frac{1}{(v - u_i)^2}, \tag{2.73}$$

in view of (2.71). Putting $p = T_n$ in (2.72) yields

$$| T_n^{(k+1)}(v) | = | T_n^{(k-1)}(v) | \sum_{i=1}^{n-(k-1)} \frac{1}{(v - u_i)^2}. \tag{2.74}$$

The induction is now completed by substituting (2.74) in (2.73). The possibility of equality is established by the induction, as well, since the inequality (2.73) is strict unless $p = T_n$.

It is not difficult to obtain a best uniform bound on $|p'(x)|$ from (2.69). Suppose that $n > 1$. If

$$|x| \leqslant \cos \frac{\pi}{2n} = \xi_1^{(n)},$$

then

$$1 - x^2 \geqslant 1 - \cos^2 \frac{\pi}{2n} = \sin^2 \frac{\pi}{2n} > \left[\frac{2}{\pi} \left(\frac{\pi}{2n} \right) \right]^2 = \frac{1}{n^2}$$

and $p \in B_n$ implies

$$|p'(x)| < n^2, \tag{2.75}$$

in view of (2.69). Next, suppose that $|x| > \xi_1^{(n)}$. If we recall (1.41), we see that

$$p'(x) = L_{n-1}(p', T; x) = \frac{T_n(x)}{n} \sum_{j=1}^{n} (-1)^{j-1} \frac{p'(\xi_j^{(n)})(1 - (\xi_j^{(n)})^2)^{1/2}}{(x - \xi_j^{(n)})}.$$

In the interval $\xi_1^{(n)} < x \leqslant 1$, $T_n(x)$ is positive as is $(x - \xi_j^{(n)})$ for $j = 1, \ldots, n$; hence, since $|p'(\xi_j^{(n)})(1 - (\xi_j^{(n)})^2)^{1/2}| \leqslant n$ [by (2.69) or, better still, (2.70)], we obtain

$$|p'(x)| \leqslant T_n(x) \sum_{j=1}^{n} \frac{1}{x - \xi_j^{(n)}} = T_n'(x). \tag{2.76}$$

Since $T_n'(x)$ is monotone increasing in $\xi_1^{(n)} \leqslant x \leqslant 1$,

$$|p'(x)| \leqslant T_n'(1) = n^2.$$

The case $-1 \leqslant x \leqslant -\xi_1^{(n)} = \xi_n^{(n)}$ is treated in the same way, and putting the pieces together yields the theorem of A. A. Markov:

If $p \in B_n$ then $\|p'\| \leqslant n^2$, with equality only for $p = \pm T_n$.

[The condition for equality stems from (2.70).] We shall have more to say about this theorem shortly.

Remark 2. $t \in \mathcal{T}_n$ and $\|t\| \leqslant 1$ imply $\|t^{(k)}\| \leqslant n^k$ by mathematical induction on the order of the derivative k.

Remark 3. Repeated application of A. A. Markov's theorem gives

$$\|p^{(k)}\| \leqslant [n(n-1) \cdots (n-(k-1))]^2, \qquad k = 1, \ldots, n,$$

a bound that is much too large. To obtain a better bound we observe that if $p \in B_n$ then for $0 < x_0 \leqslant 1$

$$q(x) = p\left[(1 + x_0) \frac{1+x}{2} - 1 \right] \in B_n,$$

and, since

$$\left| \left[\frac{(1 + x_0)}{2} \right]^k p^{(k)}(x_0) \right| = |q^{(k)}(1)| \leqslant T_n^{(k)}(1),$$

in view of (2.37), $|p^{(k)}(x_0)| \leqslant 2^k T_n^{(k)}(1)$. If $-1 \leqslant x_0 < 0$, the same result is obtained by putting

$$q(x) = p\left[(1 - x_0)\frac{x - 1}{2} + 1 \right].$$

Thus, recalling (1.97), we obtain

$$\|p^{(k)}\| \leqslant 2^k \frac{n^2(n^2 - 1) \cdots (n^2 - (k - 1)^2)}{1 \cdot 3 \cdot 5 \cdots (2k - 1)}. \tag{2.77}$$

The truth of the matter is that (2.77) remains true with the factor 2^k deleted. We establish this next, but it is no easy task.

5. *V. A. Markov's Theorem.* The direct generalization of A. A. Markov's theorem was provided by his brother, V. A. Markov [1] (see also Bernstein [3], Mohr [1], Voronovskaja [1], and Boas [1]) who showed that if $p \in B_n$ then $\|p^{(k)}\| \leqslant T_n^{(k)}(1), 0 \leqslant k \leqslant n$. We follow Duffin and Schaeffer [1] in proving the stronger result that $p \in C_n$ implies $\|p^{(k)}\| \leqslant T_n^{(k)}(1), 1 \leqslant k \leqslant n$. First let us see how far we can get with our representation formula for linear functionals. If $-1 \leqslant \xi \leqslant 1$, let

$$M_k(\xi) = \max_{p \in C_n} |p^{(k)}(\xi)|, \quad 1 \leqslant k \leqslant n.$$

We saw [(2.37)] that $M_k(1) = T_n^{(k)}(1)$. Hence $M_n(\xi) = M_n(1) = T_n^{(n)}(1) = 2^{n-1}$. Note also that

$$M_k(-\xi) = M_k(\xi), \tag{2.78}$$

for if $p_\xi^{(k)}(\xi) = M_k(\xi)$ and $q(x) = p_\xi(-x)$ then $q \in C_n, M_k(-\xi) \geqslant |q^{(k)}(-\xi)| = p_\xi^{(k)}(\xi) = M_k(\xi)$, and, reversing the roles of ξ and $-\xi$, establishes (2.78).

Suppose then that $0 \leqslant \xi \leqslant 1, 1 \leqslant k \leqslant n - 1$, and let $B = \{\eta_0^{(n)}, \ldots, \eta_n^{(n)}\}$ in Theorem 2.13. Let

$$Fp = p^{(k)}(\xi) = \sum_{j=1}^{r} \alpha_j p(y_j) \tag{2.79}$$

be a canonical representation of $p^{(k)}(\xi)$, where, of course, the y_j come from B. Since ± 1 is clearly no extremal for F, (2.79) is its unique canonical representation according to Theorem 2.15. We claim that $r \geqslant n$. If $r \leqslant n - 1$ and $\omega(x) = (x - y_1) \cdots (x - y_{n-1})$, where $y_{r+1}, \ldots, y_{n-1}$ are "new" distinct points of B, then $\omega^{(k)}(\xi) = 0$, *but* $q(x) = x\omega(x) \in \mathscr{P}_n$ and $0 = q^{(k)}(\xi) = \xi\omega^{(k)}(\xi) +$

$k\omega^{(k-1)}(\xi)$, hence $\omega^{(k-1)}(\xi) = 0$. Rolle's theorem implies that $\omega^{(k-1)}$ has only simple zeros which contradicts the existence of a zero of order at least 2 at ξ.

Now the Lagrange interpolation formula yields

$$Fp = p^{(k)}(\xi) = \sum_{i=0}^{n} p(\eta_i)\, l_i^{(k)}(\xi), \tag{2.80}$$

which is a canonical representation of F, since

$$\sum_{i=0}^{n} |\, l_i^{(k)}(\xi)\,| = M_k(\xi). \tag{2.81}$$

Thus (2.80) and (2.79) must be identical, and we have $r = n + 1$ except for the finite number of points ξ which are zeros of some $l_i^{(k)}(x)$ (keep in mind that here $B \neq I$). Note that, since $r \geqslant n$, no two $l_i^{(k)}(x)$ are zero simultaneously for any x. The fundamental polynomials $l_i^{(k)}(x)$ are given explicitly by

$$l_i(x) = \begin{cases} \dfrac{(-1)^{i+1}(1-x^2)\,T_n'(x)}{n^2(x-\eta_i)}, & i = 1, \ldots, n-1 \\[3mm] \dfrac{(-1)^{i+1}(1-x^2)\,T_n'(x)}{2n^2(x-\eta_i)}, & i = 0, n. \end{cases}$$

Therefore, if we put

$$\epsilon_i(\xi) = \operatorname{sgn} l_i^{(k)}(\xi)$$

so that (2.81) reads

$$\sum_{i=0}^{n} \epsilon_i(\xi)\, l_i^{(k)}(\xi) = M_k(\xi), \tag{2.82}$$

we see that

$$\epsilon_i(1) = (-1)^i, \qquad i = 0, \ldots, n,$$

and since $p_\xi(\eta_i) = \epsilon_i(\xi)$ we recover the fact that $p_1(x) = T_n(x)$. But, if τ denotes the largest zero of any of $l_i^{(k)}(x)$, $i = 0, \ldots, n$ (hence $-1 < \tau < 1$), then

$$\epsilon_i(\xi) = (-1)^i, \qquad i = 0, \ldots, n,$$

for $\tau < \xi < 1$ and we know that

$$p_\xi(x) = T_n(x)$$

and

$$M_k(\xi) = T_n^{(k)}(\xi), \qquad \text{for } \tau \leqslant \tau \leqslant 1. \tag{2.83}$$

We claim next that τ is the largest zero of $l_n^{(k)}(x)$. To establish this we need the following preliminary result.

LEMMA 2.7.1

If $p(x) = (x - a_1) \cdots (x - a_m)$ and $q(x) = (x - b_1) \cdots (x - b_m)$, where $b_1 > a_1 > b_2 > a_2 > \cdots > b_m > a_m$, then, if $t_1, \ldots, t_{m-1}$ are the zeros of p' and $z_1, \ldots, z_{m-1}$ are the zeros of q' (each set arranged in decreasing order), we have $z_1 > t_1 > z_2 > t_2 > \cdots > z_{m-1} > t_{m-1}$.

Proof. Since $p(x) - q(x) \in \mathscr{P}_{m-1}$, the Lagrange interpolation formula gives

$$p(x) - q(x) = \sum_{j=1}^{m} \frac{p(b_j) - q(b_j)}{q'(b_j)} \frac{q(x)}{x - b_j} = \sum_{j=1}^{m} \frac{p(b_j)}{q'(b_j)} \frac{q(x)}{x - b_j}. \qquad (2.84)$$

Suppose $q'(z) = 0$, then (2.84) implies that

$$\frac{p'(z)}{q(z)} = - \sum_{j=1}^{m} \frac{p(b_j)}{q'(b_j)} \frac{1}{(z - b_j)^2}.$$

We observe that for $j = 1, \ldots, m$, sgn $q'(b_j) = (-1)^{j-1}$ and sgn $p(b_j) = (-1)^{j-1}$. Thus for $i = 1, \ldots, m-1$,

$$\frac{p'(z_i)}{q(z_i)} < 0.$$

Now sgn $q(z_i) = (-1)^i$ and so sgn $p'(z_i) = (-1)^{i-1}$. Hence p' has exactly one zero in each interval (z_{i+1}, z_i), $i = 1, \ldots, m-2$ (for if it has more than one zero in one such interval it has at least three there), and since p' is positive at both z_1 and a_1 its remaining zero is not in (z_1, a_1). Thus p' has its remaining zero in (a_m, z_{m-1}). This establishes the lemma. ∎

Remark. It is easy to see, using mathematical induction, that the hypothesis of the lemma implies that the zeros of $p^{(k)}$ and $q^{(k)}$ for $k = 2, \ldots, m-1$ interlace in exactly the same way as those of p' and q'.

Let us consider $l_i(x)$ and $l_j(x)$ fundamental polynomials for any set of nodes $x_0 > x_1 > \cdots > x_n$, where $i > j$. There exists a nonzero constant, c, such that $l_j - cl_i \in \mathscr{P}_{n-1}$. Applying the Lagrange interpolation formula, as in the proof of Lemma 2.7.1, yields

$$\frac{l_j'(z)}{l_i(z)} = - \frac{1}{l_i'(x_i)} \frac{1}{(z - x_i)^2},$$

where z is any zero of l_i'. The leading coefficient of $l_i(x)$ has the sign of $(-1)^i$ as Exercise 1.3.6 reveals; hence

$$\text{sgn } l_i'(x_i) = (-1)^{i+j}$$

and

$$\text{sgn} \frac{l_j'(z)}{l_i(z)} = (-1)^{i+j+1}.$$

Now we can conclude, exactly as we did in the proof of Lemma 2.7.1, that the zeros of l_i' and l_j' interlace *strictly*. This fact, together with the lemma, applied to l_i' and l_j' leads us to the following conclusion.

THEOREM 2.21

Given any nodes $x_0 > x_1 > \cdots > x_n$ and $l_i(x), i = 0, \ldots, n$, the fundamental polynomials for the nodes; if $z_{i,1} > z_{i,2} > \cdots > z_{i,n-k}$ are the zeros of $l_i^{(k)}(x)$, $1 \leqslant k \leqslant n - 1$, then

$$z_{n,1} > z_{n-1,1} > \cdots > z_{0,1} > z_{n,2} > z_{n-1,2} > \cdots > z_{0,2} > \cdots >$$
$$> z_{n,n-k} > z_{n-1,n-k} > \cdots > z_{0,n-k}.$$

In particular, then, the τ in (2.83) is the largest zero of $l_n^{(k)}(x)$ or, equivalently, of $[(1-x) T_n'(x)]^{(k)} = (1-x) T_n^{(k+1)}(x) - k T_n^{(k)}(x) = ([(k-1)x - k] T_n^{(k)}(x) - [n^2 - (k-1)^2] T_n^{(k-1)}(x))/(1+x)$, in view of Exercise 1.5.5. Note that if u is the largest zero of $T_n^{(k-1)}(x)$ then certainly $\tau < u$ (cf. Exercise 1.5.10).

The choice of $x_i = \eta_i$ in Theorem 2.21 enables us to describe $M_k(\xi)$ more fully. In each interval $(z_{n-i,j}, z_{n-i+1,j}), i = 1, \ldots, n; j = 1, \ldots, n - k$ or $(z_{n,j}, z_{0,j-1}), j = 2, \ldots, n - k$, none of $\epsilon_i(\xi), i = 0, \ldots, n$, changes sign and $M_k(\xi)$ is given by the polynomial (2.82).

As we have just seen for $z_{n,1} < \xi < 1$

$$\epsilon_i(\xi) = (-1)^i, \qquad i = 0, \ldots, n.$$

If ξ now passes into the interval $(z_{n-1,1}, z_{n,1})$ the sign configuration becomes

$$\epsilon_i(\xi) = (-1)^i, \qquad i = 0, \ldots, n - 1$$
$$\epsilon_n(\xi) = (-1)^{n-1}.$$

As ξ continues to move to the left, an alteration of sign percolates through the sequence $\epsilon_n, \epsilon_{n-1}, \ldots, \epsilon_0$, until ξ passes through the point $z_{0,1}$ into $(z_{n,2}, z_{0,1})$ in which

$$\epsilon_i(\xi) = (-1)^{i-1}, \qquad i = 0, \ldots, n.$$

Thus in $(z_{n,2}, z_{0,1})$, $p_\xi(x) = -T_n(x)$ and $M_k(\xi) = -T_n^{(k)}(\xi)$. Now, when ξ moves past $z_{n,2}$, the percolation process is repeated, starting with

$$\epsilon_i(\xi) = (-1)^{i-1}, \qquad i = 0, \ldots, n - 1,$$
$$\epsilon_n(\xi) = (-1)^n,$$

for ξ in $(z_{n-1,2}, z_{n,2})$. Thus, if $p_\xi = q$ in $(z_{n-1,1}, z_{n,1})$, then $p_\xi = -q$ in $(z_{n-1,2}, z_{n,2})$, and so on.

In the case $k = n - 1$ analysis of (2.81) leads to an easy proof of the Duffin-Schaeffer-Markov result.

THEOREM 2.22

If $p \in C_n$, then for $-1 \leqslant x \leqslant 1$

$$|p^{(n-1)}(x)| \leqslant T_n^{(n-1)}(1) = 2^{n-1} n!,$$

with equality possible only if $p = \pm T_n$ and $x = \pm 1$.

Proof.   As an easy consequence of (2.82) we see that

$$l_i^{(n-1)}(x) = \begin{cases} (-1)^i \, 2^{n-1}(n-1)! \left(x + \dfrac{\eta_i}{n} \right), & i = 1, \dots, n-1, \\[3mm] (-1)^i \, 2^{n-2}(n-1)! \left(x + \dfrac{\eta_i}{n} \right), & i = 0, n. \end{cases}$$

Therefore

$$M_{n-1}(\xi) = \sum_{i=0}^{n} |l_i^{(n-1)}(\xi)| = 2^{n-1}(n-1)! \sum_{i=0}^{n} {}'' \left| \xi + \dfrac{\eta_i}{n} \right|.$$

We claim that

$$f(\xi) = \sum_{i=0}^{n} {}'' \left| \xi + \dfrac{\eta_i}{n} \right|$$

attains its maximum on $[0, 1]$ at $\xi = 1$. $f(\xi)$ is a polygonal line with possible changes in direction at $-\eta_i/n, i = 0, \dots, n$, and its slope in $-\eta_j/n < x < -\eta_{j+1}/n$, $j = 0, \dots, n-1$, is

$$\sum_{i=0}^{j} {}' 1 - \sum_{i=j+1}^{n-1} 1 - \tfrac{1}{2} = 2j + 1 - n.$$

Since only breaks corresponding to $j \geqslant [n/2]$ fall in $0 \leqslant \xi \leqslant 1$, the slopes of consecutive segments of $f(\xi)$ as ξ moves from $-\eta_{[n/2]}/n$ to 1 are nonnegative and increasing so that $f(\xi)$ is monotone increasing in $[-\eta_{[n/2]}/n, 1]$; but $-\eta_{[n/2]}/n \leqslant 0$, hence $|f(\xi)| \leqslant |f(1)|, 0 \leqslant \xi \leqslant 1$, with equality only at $\xi = 1$. Thus, if $0 \leqslant \xi < 1$,

$$M_{n-1}(\xi) < M_{n-1}(1) = T_n^{(n-1)}(1).$$

The conditions for equality follow from example 1, p. 93.   ∎

Note that $M_{n-1}(\xi)$ is a convex function of ξ on $[-1, 1]$. For k other than $n-1$ this need not be the case; indeed, it may happen that $M_k(\xi)$ is not even monotone increasing on $[0, 1]$. Let us look at some examples.

1. Suppose that $n = 2$. Clearly

$$M_1(\xi) = \begin{cases} 2\xi + 1, & 0 \leq \xi \leq \tfrac{1}{2}, \\ 4\xi, & \tfrac{1}{2} \leq \xi \leq 1. \end{cases}$$

Let us compare this with the case in which we seek to maximize $\|p'\|$ subject to $p \in B_2$ (rather than C_2). If $-1 \leq \xi \leq 1$, put

$$N_k(\xi) = \max_{p \in B_n} |p^{(k)}(\xi)|, \qquad 1 \leq k \leq n.$$

Of course, since $B_n \subset C_n$, we have $N_k(\xi) \leq M_k(\xi)$. Suppose that $n = 2$. Let us find $N_1(\xi)$. The remark immediately following Theorem 2.21, with $k = 1$ and $n = 2$, shows that

$$N_1(\xi) = T_2'(\xi) = 4\xi, \qquad \tfrac{1}{2} \leq \xi \leq 1.$$

Suppose, then, that $0 < \xi < \tfrac{1}{2}$, and

$$q_\xi'(\xi) = N_1(\xi).$$

Then for all $q \in B_2$ we have

$$q'(\xi) = \alpha_1 q(x_1) + \alpha_2 q(x_2) \tag{2.85}$$

and $q_\xi'(\xi) = |\alpha_1| + |\alpha_2|$. Both x_1 and x_2 cannot be interior points of $[-1, 1]$, for in that case $q_\xi'(x_1) = q_\xi'(x_2) = 0$. Putting $q(x) = (x - x_1)(x - x_2)$ reveals that $\xi = (x_1 + x_2)/2$, hence neither of x_1 and x_2 is -1. Suppose that $x_1 = 1$, then $x_2 = 2\xi - 1$, $q = 1$ and $q = x$ imply that $\alpha_1 + \alpha_2 = 0$, and $\alpha_1 + \alpha_2(2\xi - 1) = 1$; hence

$$\alpha_1 = \frac{1}{2(1 - \xi)}, \qquad \alpha_2 = -\frac{1}{2(1 - \xi)}$$

and $q_\xi'(\xi) = (1 - \xi)^{-1} > 4\xi = T_2'(\xi)$. The polynomial

$$q_\xi(x) = \frac{1}{2(1 - \xi)^2} (x^2 - 2(2\xi - 1)x + (2\xi^2 - 1))$$

is in B_2 for $0 \leq \xi \leq \tfrac{1}{2}$. Hence

$$N_1(\xi) = \begin{cases} \dfrac{1}{1 - \xi}, & 0 \leq \xi \leq \tfrac{1}{2}, \\ 4\xi, & \tfrac{1}{2} \leq \xi \leq 1. \end{cases} \tag{2.86}$$

Note that $M_k(\xi)$ for any n is, in view of (2.81), a piecewise polynomial function, whereas (2.86) shows that $N_1(\xi)$ for $n = 2$ is not.

2. Suppose that $n = 3$. An easy calculation yields

$$M_1(\xi) = \begin{cases} -12\xi^2 + 3; & 0 \leqslant \xi < \dfrac{\sqrt{7} - 2}{6}, \\[2ex] -\dfrac{8}{3}(3\xi^2 - \xi - 1); & \dfrac{\sqrt{7} - 2}{6} \leqslant \xi < \dfrac{\sqrt{13} - 1}{6}, \\[2ex] \dfrac{16}{3}\xi; & \dfrac{\sqrt{13} - 1}{6} \leqslant \xi < \dfrac{\sqrt{13} + 1}{6}, \\[2ex] \dfrac{8}{3}(3\xi^2 + \xi - 1); & \dfrac{\sqrt{13} + 1}{6} \leqslant \xi < \dfrac{\sqrt{7} + 2}{6}, \\[2ex] 12\xi^2 - 3; & \dfrac{\sqrt{7} + 2}{6} \leqslant \xi \leqslant 1. \end{cases} \tag{2.87}$$

By contrast Boas [1] gives

$$N_1(\xi) = \begin{cases} -12\xi^2 + 3; & 0 \leqslant \xi < \dfrac{\sqrt{7} - 2}{6}, \\[2ex] \dfrac{7\sqrt{7} + 10}{9(1 + \xi)}; & \dfrac{\sqrt{7} - 2}{6} \leqslant \xi < \dfrac{2\sqrt{7} - 1}{9}, \\[2ex] \dfrac{16\xi^3}{(9\xi^2 - 1)(1 - \xi^2)}; & \dfrac{2\sqrt{7} - 1}{9} \leqslant \xi < \dfrac{2\sqrt{7} + 1}{9}, \\[2ex] \dfrac{7\sqrt{7} - 10}{9(1 - \xi)}; & \dfrac{2\sqrt{7} + 1}{9} \leqslant \xi < \dfrac{\sqrt{7} + 2}{6}, \\[2ex] 12\xi^2 - 3; & \dfrac{\sqrt{7} + 2}{6} \leqslant \xi \leqslant 1. \end{cases}$$

The graph of $M_1(\xi)$ is shown in Figure 2.1. Observe that $M_1(\xi)$ in this case is neither convex nor monotone increasing.

A detailed characterization of $N_k(\xi)$ is found in Voronovskaja [1], where implicitly, much information about $M_k(\xi)$ can also be found. We next present Duffin and Schaeffer's improved version of V. Markov's theorem, which, surprisingly, requires an excursion into the complex plane and the elementary theory of analytic functions. The following lemmas are needed.

Fig. 2.1

LEMMA 2.7.2 (Rouché's Theorem)

If $f(z)$ and $g(z)$ are analytic inside and on a simple closed Jordan curve, C, and $|g(z)| < |f(z)|$ for all z on C then $g(z) + f(z)$ and $f(z)$ have the same number of zeros inside C.

Proof. See any text on complex function theory; e.g., Titchmarsh [1].

LEMMA 2.7.3

If all the zeros of $p(z) \in \mathscr{P}_n$ are in the half-plane $x < a$, then all the zeros of $p'(z)$ are in $x < a$.

Proof. Suppose that Re $w \geqslant a$; then $p(w) \neq 0$. Let $z_1, \ldots, z_n$ be the zeros of p, each zero appearing in the sequence according to its multiplicity. Each of the complex numbers $z_i - w$ satisfies

$$\frac{\pi}{2} < \arg (z_i - w) < \frac{3\pi}{2},$$

and the same is true of the complex numbers $(\bar{z}_i - \bar{w})^{-1} = |z_i - w|^{-2}(z_i - w)$, but then

$$\left[-\frac{p'(w)}{p(w)} \right] = \sum_{i=1}^{n} \frac{1}{\overline{z_i - w}} \neq 0,$$

since the sum of complex numbers, all of which lie in Re $z < 0$, must also lie in Re $z < 0$. Thus $p'(w) \neq 0$, and the lemma is proved. (This is a special case of Gauss's theorem: the zeros of p' lie in the convex hull of the zeros of p.) ∎

LEMMA 2.7.4

Let $a_1, \ldots, a_{2n}$ be nonnegative numbers and $a'_1, \ldots, a'_{2n}$, a rearrangement of these according to size so that $a'_1 \geqslant a'_2 \geqslant \cdots \geqslant a'_{2n} \geqslant 0$. Then for $t \geqslant 0$

$$(a_1 a_2 + t)(a_3 a_4 + t) \cdots (a_{2n-1} a_{2n} + t) \leqslant (a'_1 a'_2 + t)(a'_3 a'_4 + t) \cdots$$
$$\cdots (a'_{2n-1} a'_{2n} + t). \tag{2.88}$$

Proof. The lemma is certainly true if $n = 1$. Suppose that it is true for $n - 1$. Let

$$a'_1 = a_i; \qquad a'_2 = a_j$$

Case 1. If i is odd and $j = i + 1$ or i is even and $j = i - 1$, then either $a'_1 a'_2 + t = a_i a_{i+1} + t$ or $a'_1 a'_2 + t = a_{i-1} a_i + t$. Either equality together with the inductive hypothesis applied to the set $a_k, k = 1, \ldots, 2n; k \neq i, j$, establishes (2.88).

Case 2. i and j are not as in Case 1. Therefore, if we put

$$i' = \begin{cases} i - 1, & i \text{ even} \\ i + 1, & i \text{ odd} \end{cases}; \qquad j' = \begin{cases} j - 1, & j \text{ even,} \\ j + 1, & j \text{ odd,} \end{cases}$$

no two of the indices i, i', j, and j' coincide and $(a_i a_{i'} + t)$ and $(a_j a_{j'} + t)$ both appear among the factors on the left-hand side of (2.88). However,

$$(a'_1 a'_2 + t)(a_{i'} a_{j'} + t) - (a_i a_{i'} + t)(a_j a_{j'} + t)$$
$$= t(a'_1 - a'_j)(a'_2 - a_{i'}) \geqslant 0.$$

Thus, if $(a_i a_{i'} + t)(a_j a_{j'} + t)$ is replaced by $(a_1' a_2' + t)(a_{i'} a_{j'} + t)$, the product on the left-hand side of (2.88) does not decrease. By the inductive hypothesis the lemma holds for the set of a_k, $k = 1, \ldots, 2n$; $k \neq i, j$; hence the modified left-hand side does does not exceed the right-hand side and (2.88) is established. ∎

LEMMA 2.7.5

For each $n \geq 0$

$$|T_n(x + iy)| \leq |T_n(1 + iy)|; \qquad -1 \leq x \leq 1, \ -\infty < y < \infty.$$

Proof. Putting $x = \cos \theta$, we have

$$|T_n(x + iy)|^2 = 4^{n-1} \prod_{j=1}^{n} |x + iy - \cos \theta_j|^2$$

$$= 4^{n-1} \prod_{j=1}^{n} [(\cos \theta - \cos \theta_j)^2 + y^2]$$

$$= 4^{n-1} \prod_{j=1}^{n} \left[4 \sin^2 \frac{\theta - \theta_j}{2} \sin^2 \frac{\theta + \theta_j}{2} + y^2 \right]$$

$$= 4^{n-1} \prod_{j=1}^{n} [(1 - \cos(\theta - \theta_j))(1 - \cos(\theta + \theta_j)) + y^2]$$

$$= \frac{1}{4} \prod_{i=1}^{n} [|e^{i\theta} - e^{i\theta_j}|^2 |e^{i\theta} - e^{-i\theta_j}|^2 + 4y^2],$$

where the last step follows from the law of cosines. Let $a_{2j-1} = |e^{i\theta} - e^{i\theta_j}|^2$ and $a_{2j} = |e^{i\theta} - e^{-i\theta_j}|^2$, $j = 1, \ldots, n$, so that the sequence $a_1, a_2, \ldots, a_{2n}$ consists of the squares of the distance from a point of the unit circle, $e^{i\theta}$, to the vertices of a regular $2n$-gon inscribed in the unit circle. If θ is increased or decreased by π/n, the resulting sequence of squares of distances is a rearrangement of $a_1, \ldots, a_{2n}$. Therefore, if $\phi = \theta \pm (k\pi)/n$ satisfies

$$|\phi| \leq \frac{\pi}{2n}$$

and $b_{2j-1} = |e^{i\phi} - e^{i\theta_j}|^2$, $b_{2j} = |e^{i\phi} - e^{-i\theta_j}|^2$, $j = 1, \ldots, n$, $b_1, b_2, \ldots, b_{2n}$, is a rearrangement of $a_1, \ldots, a_{2n}$; moreover, if $\phi \geq 0$, $0 \leq b_1 \leq b_2 \leq \cdots \leq b_{2n}$, whereas, if $\phi \leq 0$, $0 \leq b_2 \leq b_1 \leq b_4 \leq b_3 \leq \cdots \leq b_{2n} \leq b_{2n-1}$. In either case

Lemma 2.7.4 yields

$$|T_n(x+iy)|^2 \leqslant \tfrac{1}{4} \prod_{j=1}^{n} [|e^{i\phi} - e^{i\theta_j}|^2 |e^{i\phi} - e^{-i\theta_j}|^2 + 4y^2]$$

$$= |T_n(\bar{x}+iy)^2|, \tag{2.89}$$

where $\bar{x} = \cos\phi$. Since $\bar{x} + iy$ lies in the strip $\xi_1 \leqslant x \leqslant 1$,

$$|\bar{x}+iy - \xi_j| \leqslant |1+iy - \xi_j|, \qquad j = 1,\ldots,n$$

(see Fig. 2.2); hence

$$|T_n(\bar{x}+iy)| \leqslant |T_n(1+iy)|,$$

and, in view of (2.89), the lemma is proved. ∎

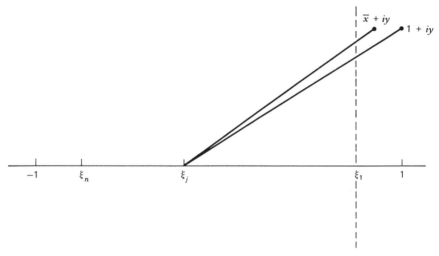

Fig. 2.2

The key to the proof of the Duffin–Schaeffer–Markov theorem is the following interesting result.

THEOREM 2.23

Let $x_1 < x_2 < \cdots < x_m < 1$ be the zeros of $v \in \mathscr{P}_m$ and suppose that

$$|v(x+iy)| \leqslant |v(1+iy)|; \qquad -1 \leqslant x \leqslant 1, -\infty < y < \infty. \tag{2.90}$$

If $q \in \mathscr{P}_m$ satisfies

$$|q'(x_i)| \leqslant |v'(x_i)|, \qquad i = 1,\ldots,m, \tag{2.91}$$

then

$$|q'(x + iy)| \leqslant |v'(1 + iy)|; \qquad -1 \leqslant x \leqslant 1, -\infty < y < \infty. \qquad (2.92)$$

Proof. Let $\xi + i\eta$ be a fixed point such that $-1 \leqslant \xi \leqslant 1$. If $v(x) = c(x - x_1) \cdots (x - x_m)$, let $r(x) = c(x - u_1) \cdots (x - u_m)$, where the zeros of r are obtained from those of v by reflecting about ξ those zeros of v that lie to the right of ξ; i.e.,

$$u_i = x_i, \qquad x_i \leqslant \xi,$$
$$u_i = 2\xi - x_i, \qquad x_i > \xi.$$

Clearly, every point $z = \xi + iy$ is equidistant from u_i and x_i, $i = 1, \ldots, m$, i.e., $|z - u_i| = |z - x_i|$; hence

$$|r(\xi + iy)| = |v(\xi + iy)|. \qquad (2.93)$$

We claim next that $|r'(\xi + i\eta)| \geqslant |q'(\xi + i\eta)|$. Put $\zeta = \xi + i\eta$. The Lagrange interpolation formula with nodes at the x_i gives

$$q'(\zeta) = v(\zeta) \sum_{i=1}^{m} \frac{q'(x_i)}{v'(x_i)} \frac{1}{\zeta - x_i}.$$

In view of (2.91), $q'(x_i)/v'(x_i) = \delta_i$, with $|\delta_i| \leqslant 1$, $i = 1, \ldots, m$. Thus, if $\eta \neq 0$,

$$|q'(\zeta)| = |v(\zeta)| \cdot \left| \sum_{i=1}^{m} \frac{\delta_i}{\xi + i\eta - x_i} \right| = |v(\zeta)| \left| \sum_{i=1}^{m} \frac{\delta_i(\xi - x_i)}{(\xi - x_i)^2 + \eta^2} \right.$$

$$\left. - i \sum_{i=1}^{m} \frac{\delta_i \eta}{(\xi - x_i)^2 + \eta^2} \right|$$

$$= |v(\zeta)| |A - iB|,$$

where

$$A = \sum_{i=1}^{m} \frac{\delta_i(\xi - x_i)}{(\xi - x_i)^2 + \eta^2}; \qquad B = \sum_{i=1}^{m} \frac{\delta_i \eta}{(\xi - x_i)^2 + \eta^2}.$$

By construction $|\xi - x_i| = \xi - u_i$; hence $(\xi - x_i)^2 = (\xi - u_i)^2$, $|\delta_i(\xi - x_i)| \leqslant \xi - u_i$, and therefore

$$|A| \leqslant \sum_{i=1}^{m} \frac{\xi - u_i}{(\xi - u_i)^2 + \eta^2} = \alpha; \qquad |B| \leqslant \sum_{i=1}^{m} \frac{|\eta|}{(\xi - u_i)^2 + \eta^2} = \beta$$

and $|A - iB| = (A^2 + B^2)^{1/2} \leqslant (\alpha^2 + \beta^2)^{1/2} = |\alpha \pm i\beta|$. By choosing the plus sign if $\eta < 0$ and the minus sign if $\eta > 0$ we obtain

$$|q'(\zeta)| \leqslant |v(\zeta)| \cdot \left| \sum_{i=1}^{m} \frac{(\xi - u_i) - i\eta}{(\xi - u_i)^2 + \eta^2} \right| = |v(\zeta)| \left| \sum_{i=1}^{m} \frac{1}{\zeta - u_i} \right|$$

$$= |v(\zeta)| \frac{|r'(\zeta)|}{|r(\zeta)|} = |r'(\zeta)|,$$

the last equality because of (2.93), thus establishing our claim for $\eta \neq 0$. The case of $\eta = 0$ now follows by continuity.

Let w be any complex number satisfying

$$|w| < 1. \tag{2.94}$$

Put $s(z) = v(z) - wr(z + \xi - 1)$. Let Γ_R be the simple closed curve consisting of the semicircle C_R: $|z - 1| = R$, Re $z \geqslant 1$, and the line segment D_R joining $1 + iR$ and $1 - iR$. If $z \in D_R$, (2.94), (2.93), and (2.90) imply that

$$|wr(z + \xi - 1)| < |r(\xi + iy)| = |v(\xi + iy)| \leqslant |v(1 + iy)| = |v(z)|.$$

Also on C_R, for R sufficiently large,

$$|wr(z + \xi - 1)| < |v(z)|,$$

since r and v have the same leading coefficient and $|w| < 1$. Thus by Rouché's theorem (Lemma 2.7.2) $s(z)$ has the same number of zeros as $v(z)$ inside Γ_R, i.e., none. Since R is arbitrary, we see that $s(z)$ has no zero in $x \geqslant 1$; hence by Lemma 2.7.3 neither has $s'(z)$. In particular, $s'(z) \neq 0$ at $z = 1 + i\eta$, i.e.,

$$v'(1 + i\eta) - wr'(\xi + i\eta) \neq 0 \tag{2.95}$$

for all $|w| < 1$. If $|v'(1 + i\eta)| < |r'(\xi + i\eta)|$, then (2.95) is violated for $w = v'(1 + i\eta)/r'(\xi + i\eta)$. Hence

$$|v'(1 + i\eta)| \geqslant |r'(\xi + i\eta)| \geqslant |q'(\xi + i\eta)|. \quad \blacksquare$$

COROLLARY 2.23.1

For $k = 1, 2, \ldots, n$

$$|T_n^{(k)}(x + iy)| \leqslant |T_n^{(k)}(1 + iy)|; \qquad -1 \leqslant x \leqslant 1, \ -\infty < y < \infty. \tag{2.96}$$

Proof. When $k = 1$, (2.96) follows from Lemma 2.7.5 and the theorem with $v = q = T_n$. If (2.96) holds for $k - 1$, then it holds for k by the theorem with $v = q = T_n^{(k-1)}$.

THEOREM 2.24 (Duffin and Schaeffer [1])

If $p \in \mathscr{P}_n$ and

$$|p(\eta_i)| \leqslant 1, \qquad i = 0, \ldots, n, \tag{2.97}$$

then for $-1 \leqslant x \leqslant 1$ and $1 \leqslant k \leqslant n$

$$|p^{(k)}(x)| \leqslant T_n^{(k)}(1) = \frac{n^2(n^2 - 1^2)(n^2 - 2^2) \cdots (n^2 - (k-1)^2)}{1 \cdot 3 \cdot 5 \cdots (2k-1)}, \tag{2.98}$$

with equality holding only if $p = \pm T_n$ and $x = \pm 1$.

Proof. Suppose $p \in C_n$ and $p \neq \pm T_n$. Then, in view of (2.71), there exists a constant $c > 1$ such that

$$|cp^{(k)}(u_i)| \leqslant |T_n^{(k)}(u_i)|, \qquad i = 1, \ldots, n - (k-1), \tag{2.99}$$

where the u_i, $i = 1, \ldots, n - (k-1)$, are the zeros of $T_n^{(k-1)}$. We now apply Theorem 2.23 with $m = n - (k-1)$, $u_i = x_i$, $v = T_n^{(k-1)}$ and $q = cp^{(k-1)}$. Observe that (2.91) holds because of (2.99) and (2.90) is simply (2.96). Thus

$$c|p^{(k)}(x + iy)| \leqslant |T_n^{(k)}(1 + iy)|, \qquad -1 \leqslant x < 1, \ -\infty < y < \infty.$$

Since $c > 1$, we have, finally,

$$|p^{(k)}(x + iy)| < |T_n^{(k)}(1 + iy)|, \qquad -1 \leqslant x \leqslant 1, \ -\infty < y < \infty. \tag{2.100}$$

Choosing $y = 0$ yields

$$|p^{(k)}(x)| < |T_n^{(k)}(1)| = T_n^{(k)}(1).$$

The observations about equality and the evaluation of $T_n^{(k)}(1)$ are consequences of Exercises 1.5.35 and 1.5.6. ∎

Remark 1.   Note that we have really proved (2.100) which is more general than (2.98).

Remark 2.   If p is a polynomial of degree at most n with complex coefficients, which satisfies (2.97), the conclusion of Theorem 2.24 still holds. To show this we note that the $l_i(x)$ with respect to the points $\eta_0, \ldots, \eta_n$ are real valued.

$$|p^{(k)}(\xi)| = \left| \sum_{i=0}^{n} p(\eta_i) l_i^{(k)}(\xi) \right| \leqslant \sum_{i=0}^{n} |l_i^{(k)}(\xi)| = \sum_{i=0}^{n} \epsilon_i(\xi) l_i^{(k)}(\xi).$$

If

$$p_\xi(x) = \sum_{i=0}^{n} \epsilon_i(\xi)\, l_i(x),$$

then $p_\xi \in C_n$, since $\epsilon_i(\xi) = \pm 1$. Thus

$$|p^{(k)}(\xi)| \leqslant p_\xi^{(k)}(\xi),$$

and the result follows from the theorem applied to p_ξ.

Remark 3.   Duffin and Schaeffer [1] also show that if E is any closed subset of $[-1, 1]$ that does not contain one of the points η_i, $i = 0, \ldots, n$, there exists $p \in \mathscr{P}_n$ satisfying

$$|p(x)| \leqslant 1, \qquad x \in E,$$

and

$$|p^{(k)}(1)| > T_n^{(k)}(1), \qquad k = 1, 2, \ldots, n.$$

There is an interesting application of Theorem 2.24 to the theory of numerical differentiation. Suppose that we wish to approximate the derivative $f^{(k)}(x)$ for $x \in I$ by $L_n^{(k)}(f, X; x)$ (cf. p. 10), $k = 1, 2, \ldots, n$. How shall we choose X? The norm of the operator

$$f \to L_n^{(k)}(f)$$

is

$$\Lambda_{n+1}^{(k)}(X) = \max_{\|f\|=1} \max_{-1 \leqslant x \leqslant 1} |L_n^{(k)}(f, X; x)| = \max_{-1 \leqslant x \leqslant 1} \sum_{j=1}^{n+1} |l_{j,n+1}^{(k)}(x)|. \tag{2.101}$$

It therefore seems desirable to choose X so that $\Lambda_{n+1}^{(k)}(X)$ is as small as possible.

In the case that $k = 0$ (2.101) is what we called (p. 12) the Lebesgue constant of order $n + 1$ of X. A set of nodes that minimizes $\Lambda_{n+1}(X)$ is not known. For $k = 1, \ldots, n$, however,

$$T_n^{(k)}(1) = \Lambda_{n+1}^{(k)}(U) \leqslant \Lambda_{n+1}^{(k)}(X), \tag{2.102}$$

where U is the array of nodes whose $(n + 1)$st row is $\eta_0^{(n)}, \ldots, \eta_n^{(n)}$. This result is due to D. L. Berman [1]. To prove (2.102) we first observe that on the one hand

$$T_n^{(k)}(1) = \left| \sum_{i=0}^{n} T_n(x_i)\, l_i^{(k)}(X; 1) \right| \leqslant \sum_{i=0}^{n} |l_i^{(k)}(X; 1)| \leqslant \Lambda_{n+1}^{(k)}(X),$$

whereas on the other Theorem 2.24 implies that

$$\sum_{i=0}^{n} |\, l_i^{(k)}(U;x)\,| \leqslant T_n^{(k)}(1), \qquad -1 \leqslant x \leqslant 1.$$

EXERCISES 2.7

2.7.1. Let

$$V = \{p \in \mathscr{P}_n / p(\pm \xi_1^{(n)}) = 0\}. \tag{2.103}$$

If $p \in V$ and

$$|\, p(x_i)\,| = \max_{-\xi_1 < x \leqslant \xi_1} |\, p(x)\,|, \qquad i = 1, \ldots, n-1,$$

where $-\xi_1 \leqslant x_1 < x_2 < \cdots < x_{n-1} \leqslant \xi_1$, then $p(x) = \pm T_n(x)$.

2.7.2. If $p \in V$ [V is defined by (2.103)],

$$\left|\, p\left(\cos\frac{j\pi}{n}\right)\right| \leqslant 1, \qquad j = 1, \ldots, n-1,$$

and $|\,t\,| \geqslant \xi_1^{(n)}$, then $|\, p^{(k)}(t)\,| \leqslant T_n^{(k)}(t)$.

2.7.3. Give an example of $p \in \mathscr{P}_2$ ($p \neq \pm T_2$) for which equality occurs in (2.45) for $j = 1$.

2.7.4. Discuss the problem of minimizing $\|\, xp' - 1\,\|$ for $p \in \mathscr{P}_n$, $p(1) = 0$, the norm being taken on the interval $[1, \alpha]$.

2.7.5. (B. Lepson [1]) When $p = T_n$, (2.69) becomes

$$|\, T_n'(x)\,| \leqslant \frac{n}{(1-x^2)^{1/2}}, \qquad -1 < x < 1. \tag{2.104}$$

Show that (2.104) can be improved to

$$|\, T_n'(x)\,| < \frac{n}{(2x)^{(n-1)/3}}, \qquad 0 < x < \tfrac{1}{2}, \, n > 1.$$

Hint. Using the 3-term recurrence formula for $R_k(t) = U_k(t/2)$ (cf. Exercise 1.5.19), show by mathematical induction that

$$|\, R_k(t)\,| \leqslant t^{-k/3}, \qquad 0 < t < 1.$$

For some other improvements on (2.104) see Askey [3].

2.7.6. Show that there exists $p \in C_n$, $p \notin B_n$, $n > 1$.

2.7.7. With the notation used in Theorem 2.21 and $x_i = \eta_i$ show that each interval $(z_{n,j}, z_{0,j-1})$, $j = 2, \ldots, n-k$; $(z_{n,1}, 1)$ and $(-1, z_{0,n-k})$ contains a zero of $T_n^{(k-1)}(x)$.

2.7.8. Show that if $p \in C_n$, then

$$|p'(x)| \leqslant T_n'(x)$$

for $x \geqslant \beta$, where $\xi_1 = \cos(\pi/2n)$ and

$$\beta = \xi_1 - \frac{1 - \xi_1^2}{\xi_1 + n^2(1 - \xi_1^2)}.$$

2.7.9. (Ehlich and Zeller [1]) Suppose that $-1 = x_1 < x_2 < \cdots < x_m = 1$ and $d_m = \max\limits_{i=1,\ldots,m-1} (x_{i+1} - x_1)$. If $p \in \mathcal{P}_n$ and $|p(x_i)| \leqslant 1$, $i = 1, \ldots, m$, then

$$\|p\| \leqslant \frac{1}{1 - (d_m^2/24)n^2(n^2 - 1)}.$$

Hint. If $\|p\| = |p(t)|$, $-1 < t < 1$ and x_j is the x_i closest to t, then

$$p(x_j) = p(t) + \frac{(x_j - t)^2}{2} p''(\theta),$$

where θ is in I. Note that $|x_j - t| \leqslant d_m/2$ and apply Theorem 2.24.

Suppose a function, $f(x)$, is sampled at points $x_i, i = 1, \ldots, n + 1$ of I with error ϵ_i at x_i. An estimate for $f(t)$, $t > 1$, can be obtained by evaluating the interpolating polynomial $p \in \mathcal{P}_n$, which satisfies $p(x_i) = f(x_i) + \epsilon_i$, at t. If

$$\epsilon = \max |\epsilon_i|, \qquad i = 1, \ldots, n + 1,$$

then the error in the extrapolation, $p(t)$, due to the ϵ_i does not exceed

$$\epsilon \sum_{i=1}^{n+1} |l_i(t)| = \epsilon \lambda_{n+1}(X; t).$$

2.7.10. Show that $\lambda_{n+1}(X; t) \geqslant \lambda_{n+1}(U; t)$, with equality only for $X = U$.

Hint. If $-1 \leqslant x_{n+1} < x_n < \cdots < x_1 \leqslant 1$, then

$$\lambda_{n+1}(U; t) = |T_n(t)| \leqslant \lambda_{n+1}(X; t).$$

2.7.11. If $-1 \leqslant x_{n+1} < x_n < \cdots < x_1 \leqslant 1$,

$$V(X) = \{p \in \mathcal{P}_n / |p(x_i)| \leqslant 1, \qquad i = 1, \ldots, n + 1\},$$

and for $t \geqslant 1, 0 \leqslant k \leqslant n$

$$\min_X \max_{p \in V(X)} |p^{(k)}(t)| = m,$$

then show that the minimal X is U and $m = T_n^{(k)}(t)$.

2.7.12. (Cavaretta [1], Matorin [1])   Let $p_n(x) = T_n(x - 1)$, the Chebyshev polynomial relative to $[0, 2]$; $p_n^{(n)}(x) = n!\, 2^{n-1}$. Let f be an n times differentiable function on $[0, \infty)$ that satisfies $\|f\| \leqslant 1$ and $\|f^{(n)}\| \leqslant n!\, 2^{n-1}$, where, if g is defined on $[0, \infty)$, we put

$$\|g\| = \sup_{0 \leqslant x < \infty} |g(x)|$$

(this notation is used in this exercise only). Show that

$$\|f^{(j)}\| \leqslant p_n^{(j)}(2) = T_n^{(j)}(1), \qquad j = 1, \ldots, n - 1. \tag{2.105}$$

Hint.   If (2.105) does not hold for some j and f, there exists $t, 0 \leqslant t < \infty$, and $a > 1$ such that $f^{(j)}(t) = a\, p_n^{(j)}(0)$.

Consider $h(x) = p_n(x) - a^{-1} f(x + t)$; h has n zeros in $[0, 2]$, hence by Rolle's theorem $h^{(i)}$ has $n - i$ zeros in $(0, 2)$.

2.7.13.   Show that the bound in (2.105) is sharp for $n = 2, 3$.

Hint.   Do some appropriate surgery on p_2 and extend it periodically.

3
EXPANSION OF FUNCTIONS IN SERIES OF CHEBYSHEV POLYNOMIALS

The Chebyshev polynomial has extremal properties in both the uniform sense (cf. Theorem 2.1) and the least squares sense [cf. (1.109)]. In Chapter 2 our main theme was extremal properties of the Chebyshev polynomials in the uniform norm. This chapter focuses on the expansion theory of Chebyshev polynomials considered as orthogonal polynomials. Particular attention is paid to the uniform approximating power of the partial sums of Chebyshev expansions.

3.1. Polynomials in Chebyshev Form

The representation of a polynomial in terms of Chebyshev polynomials (cf. Exercise 1.2.6), a particularly simple example of the expansion of a function in a series of Chebyshev polynomials, has some interesting properties; for example, let

$$p(x) = 1 + x + x^2 + x^3 + x^4 + x^5;$$ (3.1)

then, according to Exercise 1.5.32,

$$p(x) = \tfrac{15}{8} + \tfrac{19}{8} T_1(x) + T_2(x) + \tfrac{9}{16} T_3(x) + \tfrac{1}{8} T_4(x) + \tfrac{1}{16} T_5(x).$$ (3.2)

It is clear from (3.2) that

$$p_4(x) = \tfrac{15}{8} + \tfrac{19}{8} T_1(x) + T_2(x) + \tfrac{9}{16} T_3(x) + \tfrac{1}{8} T_4(x)$$ (3.3)

satisfies

$$|p(x) - p_4(x)| \leqslant \tfrac{1}{16}, \qquad -1 \leqslant x \leqslant 1,$$

and

$$p_3(x) = \tfrac{15}{8} + \tfrac{19}{8} \, T_1(x) + T_2(x) + \tfrac{9}{16} \, T_3(x) \tag{3.4}$$

satisfies

$$|p(x) - p_3(x)| \leqslant \tfrac{3}{16}, \qquad -1 \leqslant x \leqslant 1.$$

Thus (3.2) provides us with handy approximations of (3.1). We can rewrite

$$p_3(x) = \tfrac{7}{8} + \tfrac{11}{16}x + 2x^2 + \tfrac{9}{4}x^3$$

and

$$p_4(x) = 1 + \tfrac{11}{16} \, x + x^2 + \tfrac{9}{4} \, x^3 + x^4,$$

which [like the remaining partial sums of (3.2)] are called *economizations* of (3.1) (cf. Lanczos [1]), but it is by no means necessary to rewrite (3.3) and (3.4) as power polynomials, since they can be evaluated expeditiously in the Chebyshev form (3.2). Let us determine how to do this in general.

3.2. Evaluating Polynomials in Chebyshev Form

Let

$$q(x) = A_0 + A_1 \, T_1(x) + \cdots + A_n \, T_n(x), \qquad n \geqslant 2.$$

$T_n(x) = 2x \, T_{n-1}(x) - T_{n-2}(x)$ for $n \geqslant 2$; hence

$$q(x) = A_0 + A_1 \, T_1(x) + \cdots + (A_{n-2} - A_n) \, T_{n-2}(x) + (A_{n-1} + 2xA_n)$$
$$\times T_{n-1}(x)$$
$$= A_0^{(1)} + A_1^{(1)} \, T_1(x) + \cdots + A_{n-2}^{(1)} \, T_{n-2}(x) + A_{n-1}^{(1)} \, T_{n-1}(x),$$

where $A_j^{(1)} = A_j, j = 0, \ldots, n-3, A_{n-2}^{(1)} = A_{n-2} - A_n$ and $A_{n-1}^{(1)} = A_{n-1} + 2xA_n$. We now continue to apply the three-term recurrence formula to obtain the general form

$$q(x) = A_0^{(k)} + A_1^{(k)} \, T_1(x) + \cdots + A_{n-k}^{(k)} \, T_{n-k}(x)$$

by means of

$$A_j^{(k)} = A_j^{(k-1)}, \qquad j = 0, \ldots, n - (k+2),$$
$$A_{n-(k+1)}^{(k)} = A_{n-(k+1)}^{(k-1)} - A_{n-(k-1)}^{(k-1)}, \tag{3.5}$$
$$A_{n-k}^{(k)} = A_{n-k}^{(k-1)} + 2x \, A_{n-(k-1)}^{(k-1)},$$

as long as $k \leqslant n - 1$. If we put $B_k = A_{n-k}^{(k)}$, then we conclude from (3.5) that for $k = n, n - 1, \ldots, 1$,

$$B_k = 2xB_{k+1} - B_{k+2} + A_k, \tag{3.6}$$

where

$$B_{n+1} = B_{n+2} = 0. \tag{3.7}$$

When $k = n - 1$, we have

$$q(x) = A_0^{(n-1)} + A_1^{(n-1)} T_1(x)$$
$$= [A_0^{(n-2)} - A_2^{(n-2)}] + A_1^{(n-1)}x$$
$$= (A_0 - B_2) + B_1 x,$$

which yields, after defining B_0 by putting $k = 0$, in (3.6),

$$q(x) = \frac{A_0}{2} + \frac{B_0 - B_2}{2}. \tag{3.8}$$

Thus

$$p(x) = \sum_{j=0}^{n}{}' A_j T_j(x) = \frac{B_0 - B_2}{2}, \tag{3.9}$$

where B_0 and B_2 are determined by the backwards recurrence formula (3.6) with starting conditions (3.7).

In an *actual* computation using (3.6) errors will necessarily be introduced because of imprecisions in the A_k or rounding and the impossibility of doing exact arithmetic. Such errors are propagated and compounded by the recurrence. Let us attempt to estimate the resulting error in $p(x)$ (cf. Fox and Parker [1]). Suppose that ϵ_k is the local error occurring in computing B_k by (3.6), i.e., the error in B_k, assuming that B_{k+1}, B_{k+2} and A_k are correct. Let $E_m(k)$ denote the error in B_m due to exactly one unit error in the kth step $(k \geqslant m)$. Thus the total error in B_m is given by

$$\sum_{k=m}^{n} \epsilon_k E_m(k); \tag{3.10}$$

but $E_m(k)$ satisfies

$$E_m(k) = 2xE_{m+1}(k) - E_{m+2}(k) \tag{3.11}$$

with $E_{k+1}(k) = 0$ and $E_k(k) = 1$.

The three-term recurrence (3.11) has as solution

$$E_m(k) = A\, T_m(x) + B\, U_m(x),$$

and the boundary conditions then imply that

$$E_m(k) = \frac{U_{k+1}(x)\, T_m(x) - T_{k+1}(x)\, U_m(x)}{U_{k+1}(x)\, T_k(x) - U_k(x)\, T_{k+1}(x)}.$$

This last expression can be considerably simplified by using the appropriate trigonometric identities and we obtain

$$E_m(k) = U_{k-m}(x).$$

Thus, in view of (3.9) and (3.10), the error in $p(x)$ is bounded by

$$\frac{1}{2}\left[\,|\epsilon_0\,E_0(0)| + |\epsilon_1\,E_0(1)| + \sum_{k=2}^{n}\,|\epsilon_k(E_0(k) - E_2(k))|\,\right],$$

and since $E_0(k) - E_2(k) = U_k(x) - U_{k-2}(x) = 2T_k(x)$, we obtain, finally, the bound

$$\sum_{k=0}^{n}{}'\,|\epsilon_k\,|.$$

Thus the proposed method is *stable* in the sense that the resulting error is no larger than the sum of the absolute values of the local errors.

If $p(x)$ in (3.9) is an even function,

$$p(x) = \sum_{j=0}^{m}{}'\,A_{2j}\,T_{2j}(x), \tag{3.12}$$

then, since $T_{2j}(x) = T_j(T_2(x)) = T_j(2x^2 - 1)$, we need only put $t = 2x^2 - 1$ and evaluate

$$p(x) = \sum_{j=0}^{m}{}'\,D_j\,T_j(t) \tag{3.13}$$

by (3.9), where $D_j = A_{2j}, j = 0, \ldots, m$.

If $p(x)$ is given by (3.12), then

$$x\,p(x) = \sum_{j=0}^{m}{}'\,A_{2j}\,\frac{T_{|2j-1|}(x) + T_{2j+1}(x)}{2} = \sum_{j=0}^{m}\,C_{2j+1}T_{2j+1}(x),$$

where

$$C_{2j+1} = \frac{A_{2j} + A_{2j+2}}{2}, \qquad j = 0, \ldots, m$$

($A_{2m+2} = 0$). If $t = 2x^2 - 1$ and $p(x)$, written in the form (3.13) is evaluated by

$$B_k = 2tB_{k+1} - B_{k+2} + D_k; \qquad B_{m+1} = B_{m+2} = 0, \tag{3.14}$$

then

$$\sum_{j=0}^{m} C_{2j+1}\, T_{2j+1}(x) = x\, \frac{B_0 - B_2}{2}.$$ (3.15)

If, however, we consider the recurrence

$$\beta_k = 2t\beta_{k+1} - \beta_{k+2} + C_{2k+1}; \qquad \beta_{m+1} = \beta_{m+2} = 0,$$ (3.16)

which is obtained by putting $\beta_k = (B_k + B_{k+1})/2$ in (3.14), then $(B_0 - B_2)/2 = \beta_0 - \beta_1$. Thus the odd polynomial in (3.15) has the value $x(\beta_0 - \beta_1)$, which is calculable by the recurrence formula (3.16).

3.3.   Chebyshev Series

We call an infinite series of the form

$$\frac{B_0}{2} + B_1\, T_1(x) + \cdots + B_n\, T_k(x) + \cdots$$ (3.17)

a *Chebyshev series*. If

$$\sum_{k=0}^{\infty}{}' |B_k| < \infty,$$ (3.18)

then the series (3.12) is absolutely convergent for each x on I: $[-1, 1]$ and is also uniformly convergent on I (by the Weierstrass M test), so that the series (3.17) converges to a function continuous on I. If we denote the set of absolutely convergent Chebyshev series, i.e., series (3.17) satisfying (3.18), by $A(I)$ and denote the set of uniformly convergent Chebyshev series by $U(I)$, then $A(I) \subset U(I)$. However, $A(I) \neq U(I)$. We show this by the following example.

Let z be a complex variable and put

$$f_1(x) = 1 + z, \qquad g_1(z) = 1 - z.$$

We define two sequences of polynomials $f_1, f_2, \ldots,$ and $g_1, g_2, \ldots,$ by

$$
\begin{aligned}
f_{n+1} &= f_n + z^{2^n} g_n, & n = 1, 2, \ldots, \\
g_{n+1} &= f_n - z^{2^n} g_n, & n = 1, 2, \ldots ..
\end{aligned}
$$ (3.19)

It is easy to establish, by mathematical induction that f_n and g_n are polynomials of degree $2^n - 1$ with coefficients that are ± 1. Moreover, given f_n, the first 2^n coefficients of f_{n+1} are precisely the coefficients of f_n, whereas the next 2^n are obtained by recopying the first 2^{n-1} coefficients of f_n, followed by the negatives of the next 2^{n-1} coefficients of f_n. Thus we have, for example,

$$f_1(z) = 1 + z, \qquad f_2(z) = 1 + z + z^2 - z^3,$$
$$f_3(z) = 1 + z + z^2 - z^3 + z^4 + z^5 - z^6 + z^7,$$
$$f_4(z) = 1 + z + z^2 - z^3 + z^4 + z^5 - z^6 + z^7 + z^8 + z^9 + z^{10} - z^{11}$$
$$- z^{12} - z^{13} + z^{14} - z^{15}.$$

In this fashion we obtain an infinite series

$$\epsilon_0 + \epsilon_1 z + \cdots + \epsilon_k z^k + \cdots \tag{3.20}$$

with $\epsilon_k = \pm 1$ and

$$f_n(z) = \sum_{k=0}^{2^n - 1} \epsilon_k z^k.$$

This series was first considered by H. S. Shapiro [1]. We denote the partial sums of this series by

$$\mathcal{S}_k(z) = \sum_{j=0}^{k} \epsilon_j z^j, \qquad k = 1, 2, \ldots.$$

The Shapiro polynomials have the following remarkable property (cf. Rudin [1]).

LEMMA 3.3.1.

For $n = 1, 2, \ldots,$ and $0 \leqslant \theta < 2\pi$,

$$|\mathcal{S}_n(e^{i\theta})| \leqslant 5n^{1/2}. \tag{3.21}$$

Proof. We recall the identity for complex numbers

$$|\alpha + \beta|^2 + |\alpha - \beta|^2 = 2[|\alpha|^2 + |\beta|^2],$$

which yields, in view of (3.19), for $|z| = 1$, $k = 1, 2, \ldots,$

$$|f_{k+1}(z)|^2 + |g_{k+1}(z)|^2 = |f_k(z) + z^{2^k} g_k(z)|^2 + |f_k(z) - z^{2^k} g_k(z)|^2$$
$$= 2[|f_k(z)|^2 + |g_k(z)|^2].$$

Since $|f_1(z)|^2 + |g_1(z)|^2 = |1 + z|^2 + |1 - z^2| = 2^2$, we obtain for $|z| = 1$ and $k = 1, 2, \ldots,$

$$|f_k(z)|^2 + |g_k(z)|^2 = 2^{k+1}.$$

Thus, surely,

$$|f_k(e^{i\theta})| \leqslant 2^{1/2} 2^{k/2}, \tag{3.22}$$

which, since

$$f_k = \mathcal{S}_{2^k-1},$$

establishes (3.21) for $n = 2^k - 1, k = 1, 2, \ldots$.

Let $\mathcal{R}_n(z)$ be the partial sum of order $n \geq 1$ of $g_k(z)$, where $n \leq 2^k - 1$. We claim that if $1 \leq n \leq 2^k - 1$

$$|\mathcal{S}_n(e^{i\theta})| \leq (2 + 2^{1/2}) \, 2^{k/2}, \tag{3.23a}$$

$$|\mathcal{R}_n(e^{i\theta})| \leq (2 + 2^{1/2}) \, 2^{k/2}. \tag{3.23b}$$

To verify this we use mathematical induction on k. Formula 3.23 obviously holds for $k = 1$. Suppose it holds for k and suppose $1 \leq n \leq 2^{k+1} - 1$. If $n \leq 2^k - 1, |\mathcal{S}_n| = |\mathcal{R}_n| \leq (2 + 2^{1/2}) \, 2^{k/2} \leq (2 + 2^{1/2}) \, 2^{(k+1)/2}$ by the inductive hypothesis, whereas if $2^k \leq n \leq 2^{k+1} - 1$

$$|\mathcal{S}_n| \leq |f_k| + |\mathcal{R}_{n-2^k}| \leq 2^{(k+1)/2} + (2 + 2^{1/2}) \, 2^{k/2} \leq (2 + 2^{1/2}) \, 2^{(k+1)/2}$$

and

$$|\mathcal{R}_n| \leq |f_k| + |\mathcal{R}_{n-2^k}| \leq (2 + 2^{1/2}) \, 2^{(k+1)/2},$$

in view of (3.19), (3.22), and the inductive hypothesis (since $n - 2^k \leq 2^k - 1$). Thus (3.23) is established, and if $2^{k-1} \leq n \leq 2^k - 1$

$$|\mathcal{S}_n(e^{i\theta})| \leq (2 + 2^{1/2}) \, 2^{k/2} \leq 2^{1/2}(2 + 2^{1/2}) n^{1/2} \leq 5n^{1/2}. \quad \blacksquare$$

We have called this result remarkable, since any polynomial of degree n, $p(z) = a_0 + \cdots + a_n z^n$, with all coefficients of absolute value 1, satisfies

$$\frac{1}{2\pi} \int_0^{2\pi} |p(e^{i\theta})|^2 \, d\theta = |a_0|^2 + \cdots + |a_n|^2 = n + 1,$$

hence

$$\max_{0 \leq \theta < 2\pi} |p(e^{i\theta})| \geq (n + 1)^{1/2} > n^{1/2}.$$

Thus the Shapiro polynomials exhibit extremely small norm on the unit circle. With the ϵ_j as defined above we can now give our example.

THEOREM 3.1

$$\sum_{k=1}^{\infty} \frac{\epsilon_k}{k} T_k(x) \tag{3.24}$$

is uniformly convergent on I, but not absolutely convergent.

Proof. Consider

$$\sum_{k=n}^{n+m} \frac{1}{k} \epsilon_k e^{ik\theta} = \sum_{k=n}^{n+m} \frac{1}{k} (\mathscr{S}_k(e^{i\theta}) - \mathscr{S}_{k-1}(e^{i\theta}))$$

$$= \frac{\mathscr{S}_{n+m}(e^{i\theta})}{n+m} - \frac{\mathscr{S}_{n-1}(e^{i\theta})}{n}$$

$$+ \sum_{j=1}^{m-1} \mathscr{S}_{n+j}(e^{i\theta}) \frac{1}{(n+j)(n+j+1)}.$$

Then applying (3.21) and the triangle inequality yields

$$\left| \sum_{k=n}^{n+m} \frac{1}{k} \epsilon_k e^{ik\theta} \right| \leqslant \frac{10}{n^{1/2}} + 5 \sum_{j=1}^{m} \frac{1}{(n+j)^{3/2}} \leqslant \frac{10}{n^{1/2}} + 5 \sum_{k=n+1}^{\infty} \frac{1}{k^{3/2}},$$

but

$$\sum_{k=n+1}^{\infty} \frac{1}{k^{3/2}} \leqslant \int_{n}^{\infty} x^{-3/2} \, dx = \frac{2}{n^{1/2}},$$

so that

$$\left| \sum_{k=n}^{n+m} \frac{1}{k} \epsilon_k e^{ik\theta} \right| \leqslant \frac{20}{n^{1/2}}.$$

Since $|\mathrm{Re}\ z| \leqslant |z|$ for complex numbers, we have

$$\left| \sum_{k=n}^{n+m} \frac{1}{k} \epsilon_k \cos k\theta \right| \leqslant \frac{20}{n^{1/2}},$$

and putting $x = \cos \theta$ we obtain the uniform convergence of (3.24). The failure of absolute convergence is obvious, since $|\epsilon_j| = 1$ for all j. ∎

To each function $f(x)$, integrable on I, there is associated its *Chebyshev expansion*, a relationship we denote by

$$f(x) \sim \sum_{k=0}^{\infty}{}' A_k T_k(x), \qquad (3.25)$$

where

$$A_k = \frac{2}{\pi} \int_{-1}^{1} f(x) T_k(x) \frac{dx}{\sqrt{1-x^2}}, \qquad k = 0, 1, \ldots. \qquad (3.26)$$

If a Chebyshev series (3.17) converges uniformly on I and its sum is called $g(x)$, then $g(x) \in C(I)$ and the series is the Chebyshev expansion of g, for if

$$g(x) = \sideset{}{'}\sum_{k=0}^{\infty} B_k T_k(x)$$

then uniform convergence implies that

$$\frac{2}{\pi} \int_{-1}^{1} g(x) T_m(x) \frac{dx}{\sqrt{1-x^2}} = B_m, \qquad m = 0, 1, \ldots,$$

in view of the orthogonality of the Chebyshev polynomials. However, not every Chebyshev series is a Chebyshev expansion, for by the Riemann–Lebesgue lemma (cf. Zygmund [1, I, p. 45].) the coefficients in a Chebyshev expansion must satisfy $\lim_{k \to \infty} A_k = 0$.

Given $f \in C(I)$, we put

$$s_n(f; x) = s_n(x) = \sideset{}{'}\sum_{k=0}^{n} A_k T_k(x);$$

$s_n(f; x)$ is the nth partial sum of the Chebyshev expansion of f and certainly $s_n(x) \in \mathscr{P}_n$. $s_n(f)$ is a linear operator which has an explicit expression. If $x \in I$ put $x = \cos\theta, 0 \leqslant \theta \leqslant \pi$, then $f(x) = f(\cos\theta) = F(\theta)$ is defined on $[0, \pi]$ and we extend its definition to $[-\pi, 0]$ by $F(-\theta) = F(\theta)$. Thus we may consider $F(\theta)$ to be defined for all θ and have period 2π. Now

$$s_n(x) = s_n(f; \cos\theta) = \frac{1}{\pi} \sideset{}{'}\sum_{k=0}^{n} \int_{-\pi}^{\pi} F(\phi) \cos k\phi \cos k\theta \, d\phi$$

$$= \frac{1}{\pi} \int_{-\pi}^{\pi} F(\phi) \left[\sideset{}{'}\sum_{k=0}^{n} \frac{\cos k(\phi+\theta) + \cos k(\phi-\theta)}{2} \right] d\phi$$

$$= \frac{1}{\pi} \int_{-\pi}^{\pi} F(\phi) \left[\sideset{}{'}\sum_{k=0}^{n} \cos k(\phi+\theta) \right] d\phi,$$

where in the first and last steps we use the evenness of F. It is easy to verify that

$$2 \sin\frac{u}{2} \sideset{}{'}\sum_{k=0}^{n} \cos ku \equiv \sin(n + \tfrac{1}{2})u,$$

and therefore

$$s_n(x) = \frac{1}{\pi} \int\limits_{-\pi}^{\pi} F(\phi) \frac{\sin\left[(n + \frac{1}{2})(\phi + \theta)\right]}{2 \sin (\phi + \theta)/2} \, d\phi. \tag{3.27}$$

Since $s_n(f)$ is obtainable in this relatively simple manner, we may ask how well it serves as a polynomial approximation to f on I. A first observation is that $s_n(f; x)$ is the least squares approximation to f with respect to the weight function $(1 - x^2)^{-1/2}$.

THEOREM 3.2

Given $f \in C(I)$,

$$\int\limits_{-1}^{1} [f(x) - s_n(f; x)]^2 \, \frac{dx}{\sqrt{1 - x^2}} \leqslant \int\limits_{-1}^{1} [f(x) - p(x)]^2 \, \frac{dx}{\sqrt{1 - x^2}}$$

for every $p \in \mathscr{P}_n$, with equality holding only for $p = s_n(f)$.

Proof. Let $p(x) = B_0/2 + B_1 T_1(x) + \cdots + B_n T_n(x)$; then

$$\int\limits_{-1}^{1} [f(x) - p(x)]^2 \, \frac{dx}{\sqrt{1 - x^2}} = \int\limits_{-1}^{1} [f^2(x) - 2p(x)f(x) + p^2(x)] \, \frac{dx}{\sqrt{1 - x^2}}$$

$$- \int\limits_{-1}^{1} f^2(x) \, \frac{dx}{\sqrt{1 - x^2}} - \pi \sum_{k=0}^{n}{}' A_k B_k + \frac{\pi}{2} \sum_{k=0}^{n}{}' B_k^2.$$

If we use this formula for $p = s_n$ as well as for arbitrary p, we obtain

$$\int\limits_{-1}^{1} [f(x) - p(x)]^2 \, \frac{dx}{\sqrt{1 - x^2}} - \int\limits_{-1}^{1} [f(x) - s_n(x)]^2 \, \frac{dx}{\sqrt{1 - x^2}}$$

$$= \frac{\pi}{2} \sum_{k=0}^{n}{}' (B_k - A_k)^2,$$

which proves the theorem. ∎

3.4. The Relationship of S_n to E_n

We wish to investigate the relationship between $s_n(f)$ and the best *uniform* approximation to f.

If $f \in C(I)$, we put

$$S_n(f) = \|f - s_n(f)\|$$

and

$$E_n(f) = \|f - p_n^*\|,$$

where p_n^* is the best uniform approximation on I to f out of $\mathscr{P}_n$ ($\|\cdot\|$ denotes the uniform norm).

THEOREM 3.3

$$E_n(f) \leqslant S_n(f) < \left(4 + \frac{4}{\pi^2} \log n\right) E_n(f). \tag{3.28}$$

Proof. $|f - s_n(f)| = |f - p_n^* + p_n^* - s_n(f)|$
$$= |f - p_n^* + s_n(p_n^* - f)|$$
$$\leqslant E_n(f) + |s_n(p_n^* - f)|,$$

but, according to (3.27),

$$s_n(g; \cos \theta) = \frac{1}{2\pi} \int_0^{\pi} [G(\phi + \theta) + G(\phi - \theta)] \frac{\sin((2n+1)/2)\phi}{\sin(\phi/2)} d\phi, \tag{3.29}$$

and applying (3.29) with $g = p_n^* - f$ yields

$$|s_n(p_n^* - f)| \leqslant E_n(f) \cdot \frac{1}{\pi} \int_0^{\pi} \frac{|\sin((2n+1)/2)\phi|}{\sin(\phi/2)} d\phi.$$

The numbers

$$L_n = \frac{1}{\pi} \int_0^{\pi} \frac{|\sin((2n+1)/2)\phi|}{\sin(\phi/2)} d\phi$$

are known as the Lebesgue constants (of Fourier series theory) and it can be shown (cf. Rivlin [1]) that they satisfy the inequality

$$L_n < 3 + \frac{4}{\pi^2} \log n,$$

whereupon the theorem is proved. ∎

This theorem informs us that the loss in using $s_n(f)$ as a best approximation rather than p_n^* is small for an arbitrary continuous f on $[-1, 1]$. It also provides us with a convergence criterion for Chebyshev expansions if we recall Jackson's theorem (cf. Rivlin [1]); namely, if $f(x)$ is defined on $[a, b]$, we put

$$\omega(f; [a, b]; \delta) = \omega(\delta) = \sup_{\substack{x_1, x_2 \in [a, b] \\ |x_1 - x_2| \leq \delta}} |f(x_1) - f(x_2)|;$$

$\omega(\delta)$ is called the modulus of continuity of f and is defined for $\delta > 0$. It is clear that f is continuous on $[a, b]$ if, and only if, $\omega(\delta) \to 0$ as $\delta \to 0$. Jackson's theorem asserts that if $f \in C(I)$ then $E_n(f) \leq 6\omega(n^{-1})$; hence, in view of (3.28), we obtain the following.

THEOREM 3.4 (Dini-Lipschitz Test)

If $f \in C(I)$ satisfies

$$\lim_{n \to \infty} (\log n)\omega\left(\frac{1}{n}\right) = 0,$$

then $s_n(f)$ converges uniformly to f in I.

In particular, if f satisfies a Lipschitz condition of order α $(0 < \alpha \leq 1)$, its Chebyshev expansion is uniformly convergent in I.

EXERCISES 3.4

3.4.1. Verify the following Chebyshev expansions:

(a)　$\dfrac{1}{a^2 - x^2} = \dfrac{2}{a\sqrt{a^2 - 1}} \sum_{j=0}^{\infty}{}' (a - \sqrt{a^2 - 1})^{2j} T_{2j}(x)$,　　$a^2 > 1$.

(b)　$\operatorname{sgn} x = \dfrac{4}{\pi} \sum_{j-1}^{\infty} (-1)^{j-1} \dfrac{T_{2j-1}(x)}{2j - 1}$

(c)　$|x| = \dfrac{2}{\pi} + \dfrac{4}{\pi} \sum_{j=1}^{\infty} \dfrac{(-1)^{j-1}}{4j^2 - 1} T_{2j}(x)$.

(d)　$(1 - x^2)^{1/2} = \dfrac{2}{\pi} - \dfrac{4}{\pi} \sum_{j=1}^{\infty} \dfrac{1}{4j^2 - 1} T_{2j}(x)$.

(e)　$e^x = J_0(i) + 2 \sum_{k=1}^{\infty} i^k J_k(-i) T_k(x)$.

$(J_k(x)$ is the Bessel function of order k).

(f) If a, b are nonnegative integers and $a > 0$, then for $|t| < 1$

$$\frac{T_b(x) - t\, T_{b-a}(x)}{1 + t^2 - 2t\, T_a(x)} = \sum_{k=1}^{\infty} t^j\, T_{aj+b}(x).$$

(g) $\quad \dfrac{1}{x - a} = (a^2 - 1)^{-1/2} - 2(a^2 - 1)^{-1/2} \sum_{j=0}^{\infty} (a - (a^2 - 1)^{1/2})^j\, T_j(x); a > 1.$

3.4.2. (Johnson and Riess [1]) If f has an absolutely convergent Chebyshev expansion, then $L_n(f, T; x)$ converges uniformly to f on I (cf. Theorem 1.7).

Hint. Use Exercise 1.3.3 to show that

$$|f(x) - L_n(f, T; x)| \leqslant 2 \sum_{i=n+1}^{\infty} |A_i|.$$

3.4.3. Let

$$\sigma_n(f; x) = \frac{s_0(f; x) + \cdots + s_n(f; x)}{n + 1}. \tag{3.30}$$

Show that

$$\sigma_n(f; \cos \theta) = \frac{1}{\pi} \int_0^{\pi} \frac{F(\phi + \theta) + F(\phi - \theta)}{2}\, K_n(\phi)\, d\phi,$$

where

$$K_n(\phi) = \frac{1}{n + 1} \left(\frac{\sin \dfrac{n + 1}{2} \phi}{\sin \dfrac{\phi}{2}} \right)^2,$$

and $F(t) = f(\cos t)$. $K_n(\phi)$ satisfies

$$\frac{1}{\pi} \int_0^{\pi} K_n(\phi)\, d\phi = 1.$$

3.4.4. Show that

$$\sigma_n(f; x) = \sum_{j=0}^{n}{}' \left(1 - \frac{j}{n + 1} \right) A_j\, T_j(x).$$

Hence, if $f \in \mathscr{P}_n$, $\sigma_n(f) \neq f$, unless f is a constant.

3.4.5. If $m \leqslant f(x) \leqslant M$ for $x \in I$, show that

$$m \leqslant \sigma_n(f; x) \leqslant M, \qquad x \in I, \ n = 0, 1, 2, \ldots.. \tag{3.31}$$

The Fejér means defined in (3.30) have the property of staying within the bounds of the function, unlike the Fourier–Chebyshev partial sums $s_n(f)$, which may become unbounded for a bounded f. They do not, however, reproduce polynomials, which the $s_n(f)$ do. An average of the partial sums $s_n(f)$ which reproduces polynomials of appropriate degree and which remains bounded for bounded f was discovered by de La Vallée Poussin [1].

3.4.6. Let

$$\tau_{2n-1}(f; x) = \frac{s_n(f; x) + \cdots + s_{2n-1}(f; x)}{n}$$

Show that if $f \in \mathscr{P}_n$, $\tau_{2n-1}(f) = f$. Moreover, for any $f \in C(I)$,

$$|\tau_{2n-1}(f; x)| \leqslant 3\|f\|, \qquad -1 \leqslant x \leqslant 1.$$

Hint. Show that $\tau_{2n-1} = 2\sigma_{2n-1} - \sigma_{n-1}$.

3.4.7. Show that $\|f - \tau_{2n-1}(f)\| \leqslant 4E_n(f)$.

There are functions for which $S_n = E_n$; for example, if $f \in \mathscr{P}_{n+1}$, then

$$f(x) = \sum_{j=0}^{n+1}{}' A_j \, T_j(x)$$

and

$$f(x) - s_n(x) = A_{n+1} \, T_{n+1}(x).$$

Thus, if $A_{n+1} \neq 0$, in view of Theorem 2.10, $s_n = p^*$, hence $S_n = E_n$. Another class of functions for which $S_n = E_n$ is described in the following theorem.

THEOREM 3.5

If $f \in C(I)$ has the convergent Chebyshev expansion

$$\sum_{k=0}^{\infty}{}' A_k \, T_{n_k}(x) \tag{3.32}$$

with $A_k > 0$, $k = 0, 1, 2, \ldots$, then $S_n(f) = E_n(f)$ for $n = 0, 1, 2, \ldots$, if, and only if,

$$\frac{n_{k+1}}{n_k} = 2m_{k+1}, \qquad k = 0, 1, \ldots, \tag{3.33}$$

where m_k is a positive integer.

Proof.   If $n_k \leqslant n < n_{k+1}$,

$$f(x) - s_n(x) = \sum_{j=k+1}^{\infty} A_j \, T_{n_j}(x)$$

and then

$$S_n(f) = \|f - s_n\| = \sum_{j=k+1}^{\infty} A_j.$$

Suppose that (3.33) holds and

$$x_i = \eta_i^{(n_{k+1})}, \qquad i = 0, \ldots, n_{k+1}.$$

Then

$$T_{n_{k+s}}(x_i) = \cos n_{k+s} \frac{i\pi}{n_{k+1}}, \qquad s = 1, 2, \ldots,$$

and since

$$\frac{n_{k+s}}{n_{k+1}} = \frac{n_{k+s}}{n_{k+s-1}} \cdot \frac{n_{k+s-1}}{n_{k+s-2}} \cdots \frac{n_{k+2}}{n_{k+1}}$$

is an odd number by (3.33)

$$T_{n_{k+s}}(x_i) = (-1)^i, \qquad i = 0, \ldots, n_{k+1}.$$

Thus $f - s_n$ assumes the value S_n with alternating sign at $n_{k+1} + 1 \geqslant n + 2$ points of I, hence $s_n = p_n^*$ by Theorem 2.10.

For this part of the proof the requirement that (3.32) be convergent is superfluous, for the gap condition (3.33) implies the convergence of (3.32) (cf. Zygmund [1, I, Theorem 6.1]).

If $S_n = E_n$, then

$$E_n = \sum_{j=k+1}^{\infty} A_j$$

and there exist points of I, $x_0 < x_1 < \cdots < x_{n+1}$ such that for $\epsilon = 1$ or -1

$$T_{n_j}(x_i) = \epsilon(-1)^i; \qquad i = 0, \ldots, n + 1; j = k + 1, \ldots.$$

Therefore each x_i is among the $\eta_m^{(n_{k+1})}$, $m = 0, \ldots, n_{k+1}$, and $T_{n_{k+2}}(x_i) = T_{n_{k+1}}(x_i)$, so that, say,

$$\cos n_{k+2} \frac{m\pi}{n_{k+1}} = \cos m\pi,$$

which holds only if n_{k+2}/n_{k+1} is an odd integer.   ∎

If the Chebyshev expansion of f is absolutely convergent, it is clear that

$$E_n(f) \leqslant S_n(f) < |A_{n+1}| + |A_{n+2}| + \ldots. \tag{3.34}$$

Lower bounds for $E_n(f)$ in terms of the Chebyshev coefficients can also be obtained.

THEOREM 3.6

If $f \in C(I)$,

$$E_n(f) \geqslant \frac{\pi}{4} |A_{n+1}|, \tag{3.35}$$

and if the Chebyshev expansion of f is convergent

$$E_n(f) \geqslant \frac{1}{4} \left| \frac{A_{n+1} T_{n+1}(x)}{n} + \frac{2A_{n+2} T_{n+2}(x)}{n} + \cdots \right.$$

$$\left. + \frac{n-1}{n} A_{2n-1} T_{2n-1}(x) + \sum_{j=2n}^{\infty} A_j T_j(x) \right|, \qquad -1 \leqslant x \leqslant 1. \tag{3.36}$$

Proof. In view of (3.26)

$$A_{n+1} = \frac{2}{\pi} \int_{-1}^{1} f(x) T_{n+1}(x) \frac{dx}{\sqrt{1-x^2}}$$

$$= \frac{2}{\pi} \int_{-1}^{1} [f(x) - p_n^*(x)] T_{n+1}(x) \frac{dx}{\sqrt{1-x^2}};$$

hence

$$|A_{n+1}| \leqslant \frac{2}{\pi} E_n(f) \int_{-1}^{1} |T_{n+1}(x)| \frac{dx}{\sqrt{1-x^2}} = \frac{4}{\pi} E_n(f),$$

where we have used Exercise 1.5.2.

Let

$$\tau_{2n-1}(x) = \frac{s_n(x) + \cdots + s_{2n-1}(x)}{n}$$

be the de La Vallée Poussin mean of f (cf. Exercise 3.4.6). Then

$$\tau_{2n-1}(x) = s_n(x) + \left(1 - \frac{1}{n}\right)A_{n+1}\,T_{n+1}(x) + \left(1 - \frac{2}{n}\right)A_{n+2}\,T_{n+2}(x)$$

$$+ \cdots + \left(1 - \frac{n-1}{n}\right)A_{2n-1}\,T_{2n-1}(x);$$

hence

$$f(x) - \tau_{2n-1}(x) = \frac{1}{n}A_{n+1}\,T_{n+1}(x) + \frac{2}{n}A_{n+2}\,T_{n+2}(x) + \cdots$$

$$+ \frac{n-1}{n}A_{2n-1}\,T_{2n-1}(x) + \sum_{j=2n}^{\infty} A_j\,T_j(x).$$

Therefore, since $|f(x) - \tau_{2n-1}(x)| \leqslant \|f - \tau_{2n-1}\| < 4E_n(f)$, in view of Exercise 3.4.7, (3.36) is established. ∎

Remark. The bound in (3.35) cannot, in general, be improved, for, given n, consider the function sgn $T_{n+1}(x)$. Given $\epsilon > 0$, it can be "smoothed" to a continuous function $f(x)$, having zero as a best approximation and satisfying $|A_{n+1}| > (4/\pi)(1 - \epsilon)E_n(f)$.

An abundant source of lower bounds for $E_n(f)$ in terms of the Chebyshev coefficients of f is the observation that if $\rho_n(f; X)$ is the error of the best approximation on a set of $n + 2$ points X, contained in I, then $\rho_n(f; X) \leqslant E_n(f)$ (cf. the remark following Theorem 2.10).

THEOREM 3.7

If the Chebyshev expansion of f is convergent, then

$$E_n(f) \geqslant \frac{(n+2)\sin \pi/(2(n+2))}{2}\left|\sum_{m=1}^{\infty}(-1)^{m-1}(A_{(2m-1)(n+2)-1}\right.$$

$$\left. - A_{(2m-1)(n+2)+1}\right| \tag{3.37}$$

and

$$E_n(f) \geqslant |A_{n+1} + A_{3(n+1)} + A_{5(n+1)} + \cdots|. \tag{3.38}$$

Proof. Let $\xi_1, \ldots, \xi_{n+2}$ be the zeros of $T_{n+2}(x)$. Then

$$f(\xi_1, \ldots, \xi_{n+2}) = 2^{n+1} \sum_{i=1}^{n+2} \frac{f(\xi_i)}{T'_{n+2}(\xi_i)}$$

$$= \frac{2^{n+1}}{n+2} \sum_{i=1}^{n+2} (-1)^{i-1} \sin \frac{(2i-1)\pi}{2(n+2)} f(\xi_i). \tag{3.39}$$

Thus, if $g(\xi_i) = (-1)^i, i = 1, \ldots, n+2,$

$$g(\xi_1, \ldots, \xi_{n+2}) = -\frac{2^{n+1}}{n+2} \sum_{i=1}^{n+2} \sin \frac{(2i-1)\pi}{2(n+2)} = -\frac{2^{n+1}}{(n+2) \sin \pi/(2(n+2))} \tag{3.40}$$

Also, in view of (3.39), Exercises 1.2.3 and 1.2.7, and (1.141),

$$T_j(\xi_1, \ldots, \xi_{n+2}) = \begin{cases} (-1)^{m-1} 2^n; & j = (2m-1)(n+2) - 1, m = 1, 2, \ldots, \\ (-1)^m 2^n; & j = (2m-1)(n+2) + 1, m = 1, 2, \ldots, \\ 0; & \text{all other } j; \end{cases}$$

hence

$$f(\xi_1, \ldots, \xi_{n+2}) = 2^n \sum_{m=1}^{\infty} (-1)^{m-1} (A_{(2m-1)(n+2)-1} - A_{(2m-1)(n+2)+1}) \tag{3.41}$$

and

$$E_n(f) \geqslant \rho_n(f; T) = \left| \frac{f(\xi_1, \ldots, \xi_{n+2})}{g(\xi_1, \ldots, \xi_{n+2})} \right| = \frac{(n+2) \sin \pi/(2(n+2))}{2}$$

$$\times \left| \sum_{m=1}^{\infty} (-1)^{m-1} (A_{(2m-1)(n+2)-1} - A_{(2m-1)(n+2)+1}) \right|,$$

thus establishing (3.37).

As for (3.38), let $\eta_0, \ldots, \eta_{n+1}$ be the extrema of $T_{n+1}(x)$. Then [cf. (1.98)]

$$f(\eta_0, \ldots, \eta_{n+1}) = \frac{2^n}{n+1} \sum_{j=0}^{n+1} {}'' (-1)^j f(\eta_j)$$

and, if $g(\eta_i) = (-1)^{i+1}, i = 0, \ldots, n+1, g(\eta_0, \ldots, \eta_{n+1}) = -2^n$, but

$$T_j(\eta_0, \ldots, \eta_{n+1}) = \frac{2^n}{n+1} \sum_{i=0}^{n+1}{}'' T_{n+1}(\eta_i) T_j(\eta_i)$$

$$= \begin{cases} 2^n; & j = (2k-1)(n+1), \ k = 1, 2, \ldots, \\ 0; & \text{otherwise}; \end{cases}$$

hence

$$f(\eta_0, \ldots, \eta_n) = 2^n \sum_{k=1}^{\infty} A_{(2k-1)(n+1)}$$

and

$$E_n(f) \geqslant \rho_n(f; U) = \frac{|f(\eta_0, \ldots, \eta_{n+1})|}{|g(\eta_0, \ldots, \eta_{n+1})|} = |A_{n+1} + A_{3(n+1)} + \cdots|. \quad \blacksquare$$

Remark. When all of $A_{n+k}, k = 1, 2, \ldots$, are nonnegative, (3.38) is an improvement on (3.35), since it implies that

$$E_n(f) \geqslant A_{n+1}. \tag{3.42}$$

Also in this case (3.36) with $x = 1$ yields

$$E_n(f) \geqslant \frac{1}{4}\left(\frac{A_{n+1}}{n} + 2\frac{A_{n+2}}{n} + \cdots + \frac{n-1}{n}A_{2n-1} + \sum_{j=2n}^{\infty} A_j\right).$$

If the sequence of $|A_j|$ converges to zero rapidly, say so rapidly that

$$\sum_{j=2}^{\infty} |A_{n+j}| \leqslant \theta |A_{n+1}|, \tag{3.43}$$

then, in view of (3.34) and (3.35),

$$S_n(f) \leqslant (1+\theta)|A_{n+1}| \leqslant \frac{4}{\pi}(1+\theta)E_n(f)$$

and

$$1 \leqslant \frac{S_n(f)}{E_n(f)} \leqslant \frac{4}{\pi}(1+\theta), \tag{3.44}$$

so that the truncated Chebyshev expansion is "asymptotically" as good as the best approximation. Of course, if $A_{n+j} \geqslant 0, j = 1, 2, \ldots$, $4/\pi$ can be replaced by 1 in (3.44). When the function $f(x)$ is the restriction of an analytic function to I, we can estimate how fast its Chebyshev coefficients go to zero. To this end we must make a brief excursion into the complex plane.

The function

$$w(z) = \frac{(z + 1/z)}{2} \tag{3.45}$$

maps the exterior as well as the interior of $|z| = 1$ in a 1-1 conformal fashion on the whole (extended) w-plane with the interval $[-1, 1]$ deleted. Each pair of circles $|z| = \rho$, $1/\rho$ is mapped onto the same ellipse in the w-plane, C_ρ, with foci at $(\pm 1, 0)$ and the sum of major and minor axes equal to 2ρ. The mapping extends to the boundary $|z| = 1$ whose image is the interval $[-1, 1]$ (traversed twice). If Γ denotes the unit circle $|z| = 1$, then on putting $x = \cos\theta$, $0 \leqslant \theta \leqslant \pi$, in (3.26) we obtain

$$A_j = \frac{1}{\pi} \int_{-\pi}^{\pi} f(\cos\theta) \cos j\theta \, d\theta = \frac{1}{2\pi i} \int_\Gamma f\left(\frac{z + z^{-1}}{2}\right)(z^j + z^{-j}) \frac{dz}{z}. \tag{3.46}$$

THEOREM 3.8

If f is analytic inside and on the ellipse C_ρ for some $\rho (>1)$, then

$$|A_j| \leqslant \frac{2M}{\rho^j}, \tag{3.47}$$

where $M = \max |f(z)|$, $z \in C_\rho$.

Proof

$$\left| \int_\Gamma f\left(\frac{z + z^{-1}}{2}\right)(z^j + z^{-j}) \frac{dz}{z} \right| \leqslant \left| \int_\Gamma f\left(\frac{z + z^{-1}}{2}\right) z^{j-1} \, dz \right| +$$

$$+ \left| \int_\Gamma f\left(\frac{z + z^{-1}}{2}\right) z^{-j-1} \, dz \right|. \tag{3.48}$$

By Cauchy's theorem we can replace the path of integration Γ in the integrals on the right-hand side of (3.48) by $\Gamma_{\rho^{-1}} : |z| = \rho^{-1}$ and $\Gamma_\rho : |z| = \rho$, respectively, thus obtaining (3.47) in view of (3.45). ∎

COROLLARY 3.8.1

Let ρ be the largest number such that f is analytic inside C_ρ (if f is entire, $\rho = \infty$), then

$$\varlimsup_{j \to \infty} |A_j|^{1/j} \leqslant 1/\rho \tag{3.49}$$

[where the right-hand side of (3.49) is taken to be zero if $\rho = \infty$].

COROLLARY 3.8.2

If infinitely many of the Chebyshev coefficients of f are nonzero, then, given
$\epsilon > 0, \epsilon < \rho,$

$$(\rho - \epsilon)^{n+j} |A_{n+j}| < |A_n|, \qquad j = 1, 2, \ldots,$$

holds for infinitely many n.

Proof

$$(\rho - \epsilon)^{n+j} |A_{n+j}| \leqslant 2M \left(\frac{\rho - \epsilon}{\rho}\right)^{n+j},$$

and if

$$2M \left(\frac{\rho - \epsilon}{\rho}\right)^{n+j} \geqslant |A_n|, \qquad j = 1, 2, \ldots,$$

for all $n > N$ then $A_n = 0$ for all $n > N$.

Corollary 3.8.1 suggests that ellipses C_ρ might play the role of circles in the theory of Taylor series for Chebyshev expansions of analytic functions. That is indeed the case as we shall see next.

LEMMA 3.4.1

If $p \in \mathscr{P}_n$ satisfies $|p(x)| \leqslant M$ on $-1 \leqslant x \leqslant 1$, then $|p(z)| \leqslant M \rho^n$ on C_ρ for each $\rho \geqslant 1$.

Proof.   Suppose $w \in C_\rho$; then $p(w) = p((z + z^{-1})/2)$, where $|z| = \rho$, but

$$\frac{p[(z + z^{-1})/2]}{z^n}$$

is analytic in $|z| \geqslant 1$, and so the maximum of its modulus on $|z| = \rho$ cannot exceed its maximum modulus on $|z| = 1$ (the boundary).   ■

THEOREM 3.9

If, for some $\rho > 1$,

$$\overline{\lim_{n \to \infty}} \ [E_n(f)]^{1/n} \leqslant \frac{1}{\rho},$$

then f is analytic inside the ellipse C_ρ.

Proof. Suppose that $\rho_0 < \rho$ so that $E_n(f) \leqslant \rho_0^{-n}$ for $n > N$. Let p_n be the best approximation to f out of $\mathscr{P}_n$ on I, $n = 0, 1, 2, \ldots$. Thus

$$f = p_0 + (p_1 - p_0) + \cdots + (p_{n+1} - p_n) + \cdots \tag{3.50}$$

uniformly on I and on I (for $n > N$),

$$|p_{n+1} - p_n| = |f - p_n + p_{n+1} - f| \leqslant E_n(f) + E_{n+1}(f)$$
$$\leqslant \rho_0^{-n} + \rho_0^{-(n+1)} \leqslant 2\rho_0^{-n}.$$

Lemma 3.4.1 now implies that

$$|p_{n+1}(z) - p_n(z)| \leqslant 2\rho_0^{-n} \rho_1^{n+1} = 2\rho_1 \left(\frac{\rho_1}{\rho_0}\right)^n,$$

on the ellipse C_{ρ_1}. Choose $\rho_1 < \rho_0$; then the series in (3.50) is uniformly convergent inside and on C_{ρ_1}, hence f is analytic, and since ρ_1 may be chosen arbitrarily close to ρ the theorem is proved. ∎

THEOREM 3.10

The following are equivalent:

(i) f *is analytic inside the ellipse* C_ρ *with* $\rho > 1$, *but not inside* $C_{\rho'}$ *with*
    $\rho' > \rho$,

(ii) $\overline{\lim_{n \to \infty}} \ [E_n(f)]^{1/n} = \dfrac{1}{\rho}, \qquad \rho > 1,$

(iii) $\overline{\lim_{n \to \infty}} \ |A_n|^{1/n} = \dfrac{1}{\rho}, \qquad \rho > 1$

    (*with the usual interpretations if* $\rho - \infty$).

Proof. We shall prove that (iii) $\Rightarrow$ (ii) $\Rightarrow$ (i) $\Rightarrow$ (iii).

(iii) $\Rightarrow$ (ii):

In view of (3.35)

$$\overline{\lim} \ E_n^{1/n} \geqslant \overline{\lim} \ \left[\left(\frac{\pi}{4}\right)^{1/n} (|A_{n+1}|^{1/n+1})^{(n+1)/n}\right],$$

so that

$$\overline{\lim} \ E_n^{1/n} \geqslant \frac{1}{\rho}.$$

If $1 < \rho' < \rho$, then for n sufficiently large $|A_{n+1}| + |A_{n+2}| + \cdots \leqslant (\rho')^{-n}(\rho' - 1)^{-1}$; hence, in view of (3.34),

$$\overline{\lim} \, E_n^{1/n} \leqslant \frac{1}{\rho'}, \qquad \text{each } \rho' < \rho.$$

(ii) is established.

(ii) $\Rightarrow$ (i):

According to Theorem 3.9, f is analytic inside C_ρ. If f is analytic inside $C_{\rho'}$, $\rho' > \rho$, then by Corollary 3.8.1

$$\overline{\lim} \, |A_n|^{1/n} \leqslant \frac{1}{\rho'} < \frac{1}{\rho},$$

which contradicts (ii), since we have shown that (iii) $\Rightarrow$ (ii).

(i) $\Rightarrow$ (iii):

Again by Corollary 3.8.1

$$\overline{\lim} \, |A_n|^{1/n} \leqslant \frac{1}{\rho}.$$

If the inequality holds, then

$$\overline{\lim} \, E_n^{1/n} = \overline{\lim} \, |A_n|^{1/n} = \frac{1}{\rho'}, \qquad \rho' > \rho,$$

implies that f is analytic inside $C_{\rho'}$, contradicting (i). ∎

We shall call ρ, as defined in Theorem 3.10, the index of convergence of f. The index of convergence of an entire function is infinite. If f is not a polynomial, has the index of convergence $\rho < \infty$, and $0 < \epsilon < \rho - 1$, then in view of Corollary 3.8.2

$$\sum_{j=2}^{\infty} |A_{n+j}| \leqslant \frac{1}{\rho - \epsilon - 1} |A_{n+1}| \tag{3.51}$$

for infinitely many n, so that according to (3.44)

$$1 \leqslant \frac{S_n(f)}{E_n(f)} \leqslant \frac{4}{\pi} \frac{\rho - \epsilon}{\rho - \epsilon - 1}, \tag{3.52}$$

for infinitely many n. Also, since

$$|A_{n+1} + A_{3(n+1)} + \cdots| \geqslant |A_{n+1}| - \sum_{j=2}^{\infty} |A_{n+j}| \geqslant \left(1 - \frac{1}{\rho - \epsilon - 1}\right) |A_{n+1}|$$

and

$$\sum_{j=1}^{\infty} |A_{n+j}| \leqslant \left(1 + \frac{1}{\rho - \epsilon - 1}\right) |A_{n+1}|,$$

we find, in view of (3.34) and (3.38), that

$$\left(1 - \frac{1}{\rho - \epsilon - 1}\right) |A_{n+1}| \leqslant E_n(f) \leqslant S_n(f) \leqslant \left(1 + \frac{1}{\rho - \epsilon - 1}\right) |A_{n+1}|$$

$$(3.53)$$

holds for infinitely many n.

If f is an entire function, then as a consequence of (3.53) we have a result of S. Bernstein [2],

$$E_n(f) \sim |A_{n+1}| \qquad (3.54)$$

as n goes to infinity through some sequence of integers. Whether (3.54) can hold for a function that is not entire is not known.

EXERCISES 3.4 (continued)

3.4.8. With

$$\eta_j = \cos \frac{j\pi}{k+1}, \qquad j = 1, \ldots, k,$$

show that if $g(\eta_j) = (-1)^j, j = 1, \ldots, k,$

$$\frac{|f(\eta_1, \ldots, \eta_k)|}{|g(\eta_1, \ldots, \eta_k)|} = \left| \sum_{j=0}^{\infty} \Delta_2^2 A_{(2j+1)(k+1)-2} \right|,$$

where $\Delta_2^2 A_m = A_m - 2A_{m+2} + A_{m+4}$.

3.4.9. Show that

$$E_n(f) \geqslant \left| \sum_{j=0}^{\infty} \Delta_2^2 A_{(2j+1)(n+3)-2} \right|.$$

3.4.10. Show that

$$S_n(|x|) = \frac{2}{\pi(2[n/2]+1)}.$$

Hint. Exercise 3.4.1c.

3.4.11. Show that for $n > 1$

$$E_n(|x|) \geqslant \frac{1}{2\pi(2n-1)}.$$

Hint.   Put $x = 0$ in (3.36) and use Exercise 3.4.10.

3.4.12. Show that

$$\frac{1}{4\pi} \leqslant n\, E_n(|x|) \leqslant \frac{2}{\pi}.$$

S. N. Bernstein [1] showed that

$$\lim_{n \to \infty} 2n\, E_{2n}(|x|) = c = .282 \ldots .$$

It is not known whether $c = 1/(2\pi^{1/2}) = .282 \ldots .$

3.5.   The Evaluation of Chebyshev Coefficients

The usefulness of the partial sums $s_n(f)$ of the Chebyshev expansion of f has been established in the preceding section and so it becomes important to examine methods of obtaining these partial sums, i.e., methods of evaluating the coefficients $A_k = A_k(f)$ defined in (3.26). Since these coefficients are defined by integrals, our approach is to examine various methods of quadrature, or numerical integration, as applied to (3.26).

Our first approach is to apply the Gaussian quadrature formula given by (1.123) to (3.26). If we use the formula based on $\xi_1^{(m)}, \ldots, \xi_m^{(m)}$, we obtain as an approximation to A_k

$$\alpha_k^{(m)} = \frac{2}{m} \sum_{i=1}^{m} f(\xi_i^{(m)})\, T_k(\xi_i^{(m)}), \tag{3.55}$$

and since the quadrature formula is exact for $fT_k \in \mathscr{P}_{2m-1}$ we have

$$\alpha_k^{(m)} = A_k(f) \tag{3.56}$$

for $f \in \mathscr{P}_{2m-1-k}$.

Let us observe at once that if $0 \leqslant k \leqslant 2m$ (3.55) can be rewritten in the form

$$\alpha_k^{(m)} = \frac{2}{m} \sum_{i=1}^{m} f(\xi_i^{(m)})\, T_{2i-1}(\eta_k^{(2m)}), \tag{3.57}$$

since

$$T_k(\xi_i^{(m)}) = T_k\left(\cos(2i-1)\frac{\pi}{2m}\right) = T_k\left(T_{2i-1}\left(\cos\frac{\pi}{2m}\right)\right)$$

$$= T_{2i-1}\left(T_k\left(\cos\frac{\pi}{2m}\right)\right) = T_{2i-1}\left(\cos\frac{k\pi}{2m}\right)$$

in view of $T_m(T_n(x)) = T_n(T_m(x))$ (cf. Exercise 1.1.6).

The advantage of the form (3.57) is that $\alpha_k^{(m)}$ can be evaluated by using the recurrence formula (3.16).

If the Chebyshev expansion of f converges to f in $[-1, 1]$ and $m > k$,

$$\alpha_k^{(m)} = \frac{2}{m} \sum_{i=1}^{m}\left(\sum_{j=0}^{\infty}{}' A_j T_j(\xi_i^{(m)})\right) T_k(\xi_i^{(m)})$$

$$= \sum_{j=1}^{\infty}{}' A_j\left(\frac{2}{m}\sum_{i=1}^{m} T_j(\xi_i^{(m)}) T_k(\xi_i^{(m)})\right)$$

$$= A_k + \sum_{j=1}^{\infty} (-1)^j (A_{2jm-k} + A_{2jm+k}) \tag{3.58}$$

in view of (1.141). Note that if we put $m = k$ in (3.58) we obtain $\alpha_k^{(k)} = 0$, which is, of course, obvious from (3.55). Formula 3.58 provides an estimate for the error $A_k - \alpha_k^{(m)}$.

Let us put

$$u_n^{(m)}(x) = \sum_{k=0}^{n}{}' \alpha_k^{(m)} T_k(x), \tag{3.59}$$

which is the approximate value of $s_n(x)$ obtained by using the approximation (3.55) to the A_k. The polynomials $u_n^{(m)}(x)$ are also related to interpolating polynomials.

THEOREM 3.11

If $n < m$,

$$u_n^{(m)}(x) = s_n(L_{m-1}(f, T); x). \tag{3.60}$$

Proof. Let

$$L_{m-1}(x) = \sum_{j=0}^{m-1}{}' b_j T_j(x)$$

so that

$$b_j = \frac{2}{\pi} \int_{-1}^{1} L_{m-1}(x) T_j(x) \frac{dx}{\sqrt{1-x^2}}, \qquad j = 0, \ldots, m-1.$$

Thus, according to (3.56), we have

$$b_j = \frac{2}{m} \sum_{i=1}^{m} L_{m-1}(\xi_i^{(m)}) T_j(\xi_i^{(m)})$$

$$= \frac{2}{m} \sum_{i=1}^{m} f(\xi_i^{(m)}) T_j(\xi_i^{(m)}) = \alpha_j^{(m)}$$

for $j = 0, \ldots, m-1$. Since $m - 1 \geq n$, (3.60) follows. ■

Remark. If $m = n + 1$, then $u_n^{(n+1)}(x) = L_n(f, T; x)$, i.e., $u_n^{(n+1)}$ is the polynomial that interpolates f at the zeros of T_{n+1}.

Since, for $-1 \leq x \leq 1$ and $n < m$,

$$| s_n(x) - u_n^{(m)}(x) | = \left| \sum_{k=0}^{n}{}' (A_k - \alpha_k^{(m)}) T_k(x) \right|$$

$$= \left| \sum_{k=0}^{n}{}' \left(\sum_{j=1}^{\infty} (-1)^{j-1}(A_{2jm-k} + A_{2jm+k}) T_k(x) \right) \right|$$

$$\leq \sum_{j=1}^{\infty} \sum_{k=0}^{n}{}' (|A_{2jm-k}| + |A_{2jm+k}|)$$

$$\leq \sum_{j=1}^{\infty} \sum_{i=2jm-n}^{2jm+n} |A_i|,$$

we obtain

THEOREM 3.12

If $m > n$,

$$\| s_n - u_n^{(m)} \| \leq \sum_{j=1}^{\infty} \sum_{i=2jm-n}^{2jm+n} |A_i|. \qquad (3.61)$$

Our second approach is to use the integration formula of Lobatto-Markov [cf. (1.145), Exercise 1.5.29],

$$\int_{-1}^{1} g(x) \frac{dx}{\sqrt{1-x^2}} = \frac{\pi}{n} \sum_{i=0}^{n} {}'' g(\eta_i^{(n)}), \qquad g \in \mathscr{P}_{2n-1}, \tag{3.62}$$

to evaluate $A_k(f)$.

Putting $g = fT_k$ in (3.62) leads to an approximation to A_k,

$$\beta_k^{(m)} = \frac{2}{m} \sum_{i=0}^{m} {}'' T_k(\eta_i^{(m)}) f(\eta_i^{(m)}), \tag{3.63}$$

and since the quadrature formula is exact for $fT_k \in \mathscr{P}_{2m-1}$ we have

$$\beta_k^{(m)} = A_k(f) \tag{3.64}$$

for $f \in \mathscr{P}_{2m-1-k}$.

We observe, once again, that if $0 \le k \le m$ (3.63) can be written in the form

$$\beta_k^{(m)} = \frac{2}{m} \sum_{i=0}^{m} {}'' f(\eta_i^{(m)}) T_i(\eta_k^{(m)}),$$

which allows $\beta_k^{(m)}$ to be evaluated by means of the recurrence formula (3.6).

If the Chebyshev expansion of f converges to f in $[-1, 1]$ and we replace f by its expansion in (3.63), we obtain, recalling (1.144), for $m \ge n$

$$\beta_k^{(m)} = A_k + \sum_{j=1}^{\infty} (A_{2jm-k} + A_{2jm+k}). \tag{3.65}$$

Let us choose $m \ge n$ and consider

$$v_n^{(m)}(x) = \sum_{k=0}^{n} {}'' \beta_k^{(m)} T_k(x) \tag{3.66}$$

as an approximation to $s_n(x)$; $v_n^{(n)}$ is an interpolating polynomial to f.

THEOREM 3.13

If $n > 0$,

$$v_n^{(n)}(x) = L_n(f, U; x). \tag{3.67}$$

Proof.   Let

$$L_n(f, U; x) = \sum_{j=0}^{n} c_j T_j(x),$$

so that

$$c_j = \frac{2}{\pi} \int_{-1}^{1} L_n(x) \, T_j(x) \frac{dx}{\sqrt{1-x^2}}, \qquad j = 0, \ldots, n-1.$$

According to (3.64), we have, for $j = 0, \ldots, n-1$,

$$c_j = \frac{2}{n} \sum_{i=0}^{n}{''} L_n(\eta_i^{(n)}) \, T_j(\eta_i^{(n)})$$

$$= \frac{2}{n} \sum_{i=0}^{n}{''} f(\eta_i^{(n)}) \, T_j(\eta_i^{(n)}) = \beta_j^{(n)}.$$

Equation 3.65 implies that $\beta_n^{(n)} = 2c_n$ and so (3.67) is established. ■
   Comparing (3.65) with (3.58) suggests that

$$\gamma_k^{(m)} = \frac{\alpha_k^{(m)} + \beta_k^{(m)}}{2} \tag{3.68}$$

is a more useful approximation to A_k than either $\alpha_k^{(m)}$ or $\beta_k^{(m)}$, since

$$\gamma_k^{(m)} = A_k + \sum_{j=1}^{\infty} (A_{4jm-k} + A_{4jm+k}). \tag{3.69}$$

Put

$$w_n^{(m)}(x) = \sum_{k=0}^{n} \gamma_k^{(m)} \, T_k(x);$$

we then have Theorem 3.14 as an immediate consequence of (3.69).

THEOREM 3.14

If $m \geqslant n$,

$$\| s_n - w_n^{(m)} \| \leqslant \sum_{j=1}^{\infty} \sum_{i=0}^{2n} |A_{4jm-n+i}| \leqslant \sum_{i=0}^{\infty} |A_{4m-n+i}|.$$

Note that if we choose $m = n$, in which case $\alpha_n^{(n)} = 0$, we obtain

$$\| s_n - w_n^{(n)} \| \leqslant \sum_{i=0}^{\infty} |A_{3n+i}|.$$

When $k = m \neq 0$ (3.63) takes on a particularly simple form,

$$\beta_k^{(k)} = \frac{2}{k} \sum_{i=0}^{k} {}''(-1)^i f(\eta_i^{(k)}).$$ (3.70)

However, $\beta_k^{(k)} = A_k(f)$ only for $f \in \mathscr{P}_{k-1}$, when both are zero. It turns out that we can get equality for the largest possible class of polynomials by simply using $\gamma_k^{(k)} = \beta_k^{(k)}/2$ ($\alpha_k^{(k)} = 0$).

THEOREM 3.15

The quadrature formula

$$A_k(f) = \frac{\beta_k^{(k)}}{2} = \gamma_k^{(k)} = \frac{1}{k} \sum_{i=0}^{k} {}''(-1)^i f(\eta_i^{(k)})$$ (3.71)

is valid for $f \in \mathscr{P}_{3k-1}, k = 1, 2, \dots$. Moreover, when $k > 1$, there is no quadrature formula

$$A_k(f) = \sum_{i=0}^{k} c_i f(x_i)$$ (3.72)

holding for all $f \in \mathscr{P}_{3k}$, and (3.71) is the only formula of type (3.72) valid for $f \in \mathscr{P}_{3k-1}$.

Proof. The validity of (3.71) for $f \in \mathscr{P}_{3k-1}$ follows from (3.69) by choosing $m = k$.

Suppose (3.72) holds for $f \in \mathscr{P}_{3k-1}$. Put $\omega(x) = (x - x_0) \cdots (x - x_k)$ and consider $f(x) = (1 - x^2)T_k'(x)\omega(x)/(x - x_i) \in \mathscr{P}_{2k+1}$. Note that, since $k \geqslant 2$, $2k + 1 \leqslant 3k - 1$. Equations 3.72 and 3.71 imply that $0 = A_k(f) = c_i f(x_i) = c_i(1 - x_i^2) T_k'(x_i) \omega'(x_i)$. Since $c_i \omega'(x_i) \neq 0$, x_i must be a zero of $(1 - x^2)T_k'(x)$, and this must be true for $i = 0, \dots, k$. Taking $f(x) = \omega(x)/(x - x_i)$ in (3.71) and (3.72), we obtain the uniqueness of (3.71). Finally (3.71) does not hold for $f(x) = (1 - x^2)[T_k'(x)]^2 T_k(x) \in \mathscr{P}_{3k}$. ∎

Remark 1. (3.71) may be written as

$$A_k(f) = 2^{1-n} f(\eta_0^{(k)}, \dots, \eta_k^{(k)}).$$ (3.73)

Remark 2. When $k = 1$, we can improve on (3.71). There is a unique formula of the form (3.72) valid for $f \in \mathscr{P}_4$ but none valid for $f \in \mathscr{P}_5$. This highest degree of precision formula is given by $x_0 = \sqrt{\frac{3}{2}}, x_1 = -\sqrt{\frac{3}{2}}, c_0 = \frac{2}{3}, c_1 = -\frac{2}{3}$, as the reader may readily verify.

Consider the quadrature formula

$$A_k(f) = \sum_{i=0}^{l} c_i f(x_i). \tag{3.74}$$

Let $h(l)$ be the largest integer such that there exists a formula (3.74) valid for $f \in \mathscr{P}_h$. We say such a formula is of the *highest degree of precision*. We have just seen that, for $k > 1$, $h(k) = 3k - 1$. Let us examine the behavior of $h(l)$ as l varies.

THEOREM 3.16

1. $h(l) = l - 1,$ $l = 1, \ldots, k - 1.$
2. *For* $m = 1, 2, \ldots$,
    (a) $h((2m - 1)k) = (4m - 1)k - 1,$ $k > 1,$
    (b) $h(2mk - 1) = (4m + 1)k - 1.$
3. (a)   If $(2mk - 1) < l < (2m + 1)k,$ $m = 1, 2, \ldots,$
           $h(l) < 2mk + l.$
    (b)   If $(2m - 1)k < l < 2mk - 1,$ $m = 1, 2, \ldots,$
           $h(l) < (2m - 1)k + l + 1.$

Proof. 1.   If $h(l) \geqslant l$ and $1 \leqslant l \leqslant k - 1$, then, in view of the orthogonality property of the Chebyshev polynomials,

$$0 = A_k(f) = \sum_{i=0}^{l} c_i f(x_i)$$

for $f \in \mathscr{P}_l$. Put $\omega(x) = (x - x_0) \cdots (x - x_l)$ and $f = \omega(x)/(x - x_i) \in \mathscr{P}_l$; then $c_i = 0$, which is a contradiction. Thus $h(l) = l - 1$. Given any set of nodes, $x_0, \ldots, x_l$, the l linear homogeneous equations in $l + 1$ unknowns

$$\sum_{i=0}^{l} c_i x_i^j = 0, \qquad j = 0, \ldots, l - 1,$$

have a nontrivial solution and, indeed, one for which no $c_i = 0$, for if, say, $c_q = 0$, the system

$$\sum_{\substack{i=0 \\ i \neq q}}^{l} c_i x_i^j = 0, \qquad j = 0, \ldots, l - 1,$$

has only the trivial (zero) solution, since its matrix is the Vandermonde type. Since $A_k(f) = 0$ for $f \in \mathscr{P}_{l-1}$, we have $h(l) = l - 1$.

2a. Replace m by $2mk$ and g by fT_k in (3.62). Since each zero of T_k is found among the $\eta_j^{(2mk)}$, we obtain from (3.62)

$$A_k(f) = \sum_{j=0}^{(2m-1)k} b_j f(\lambda_j) \tag{3.75}$$

(where the λ_j are the η_j that are not zeros of T_k) valid for $f \in \mathscr{P}_{(4m-1)k-1}$. If there is a formula (3.74) valid for $f \in \mathscr{P}_{(4m-1)k}$, then we put $\Omega(x) = (x - \lambda_0) \cdots (x - \lambda_{(2m-1)k})$ and $\omega(x) = (x - x_0) \cdots (x - x_{(2m-1)k})$. Thus

$$f(x) = \Omega(x) \frac{\omega(x)}{x - x_i} \in \mathscr{P}_{(4m-1)k-1},$$

since $k > 1$. Hence $A_k(f) = 0$ by (3.75) and therefore

$$c_i \, \omega'(x_i) \, \Omega(x_i) = 0,$$

which implies that $\Omega = \omega$, but (3.75) cannot be exact for $f \in \mathscr{P}_{(4m-1)k}$, as the choice

$$f(x) = (1 - x^2)[(x - \lambda_1) \cdots (x - \lambda_{(2m-1)k-1})]^2 T_k(x) \in \mathscr{P}_{(4m-1)k}$$

demonstrates.

2b. $h(2mk - 1) < (4m + 1)k$, since $\omega^2 T_k \in \mathscr{P}_{(4m+1)n}$. By Gaussian quadrature [cf. (1.108)]

$$\int_{-1}^{1} g(x) \frac{dx}{\sqrt{1 - x^2}} = \frac{\pi}{(2m + 1)k} \sum_{j=1}^{(2m+1)k} g(\xi_j^{((2m+1)k)}), \tag{3.76}$$

is exact for $g \subset \mathscr{P}_{2(2m+1)n-1}$. Hence

$$A_k(f) = \sum_{j=0}^{2mk-1} d_j f(\mu_j), \tag{3.77}$$

where the μ_j are the zeros of $T_{(2m+1)k}$ which are not also zeros of T_k, is exact for $f \in \mathscr{P}_{(4m+1)k-1}$.

3a. Equation 3.74 holds for $f \in \mathscr{P}_{h(l)}$. Let $\Omega(x) = (x - \mu_0) \cdots (x - \mu_{2mk-1}) \in \mathscr{P}_{2mk}$ and $\omega(x) = (x - x_0) \cdots (x - x_l) \in \mathscr{P}_{l+1}$, then

$$f(x) = \Omega(x) \frac{\omega(x)}{x - x_1} \in \mathscr{P}_{2mk+l},$$

and in view of (3.77) we conclude in a now familiar fashion that $h(l) < 2mk + l$.

3b. The argument resembles 3a and we omit it. ∎

Remark. This theorem has the surprising implication that the addition of nodes to a quadrature formula may result in *reducing* the highest degree of precision.

The last two theorems are more interesting when we seek approximations of individual A_k rather than of s_n, since they require a different set of nodes for each k.

EXERCISE 3.5

3.5.1. Show that

$$u_n^{(n)}(x) = L_{n-1}(f, T; x).$$

3.5.2. Show that

$$\| f - u_n^{(n+1)} \| \leqslant 2 \sum_{i=n+1}^{\infty} |A_i| - \sum_{i=1}^{\infty} |A_{(2i-1)(n+1)}|.$$

3.5.3. Let

$$p_{n-1}(x) = \sum_{k=0}^{n-1}{}' \beta_k^{(n)} T_k(x).$$

Show that p_{n-1} is the best approximation to f on $\{\eta_0^{(n)}, \ldots, \eta_n^{(n)}\}$.

Hint. $p_{n-1} = v_n^{(n)} - (\beta_n^{(n)}/2) T_n$; recall Exercise 2.4.26.

3.5.4. Show that

$$\| f - w_n^{(n)} \| \leqslant 2 \sum_{i=n+1}^{\infty} |A_i| - \sum_{i=n+1}^{3n-1} |A_i|.$$

3.6. An Optimal Property of Chebyshev Expansions

As mentioned in Chapter 1, Section 1.5, the Chebyshev polynomials are members of larger families of sets of orthogonal polynomials. Each integrable function has an associated expansion in orthogonal polynomials of these other sets as well. We want to show next that in certain cases the Chebyshev expansion is best.

Let us recall that the ultraspherical polynomials $\{p_n^{(\lambda)}(x)\}_{n=0}^{\infty}$ consist of polynomials ($p_n^{(\lambda)}$ being of degree n) orthogonal on I with respect to the weight function $w_\lambda(x) = (1 - x^2)^{\lambda - 1/2}$, where $\lambda > -(\frac{1}{2})$. Thus, if we normalize the ultraspherical polynomials so that $p_n^{(\lambda)}(1) = 1$, then $p_n^{(0)} = T_n$, and the case $\lambda = \frac{1}{2}$ gives the Legendre polynomials, whereas $\lambda = 1$ corresponds to the Chebyshev polynomials of the second kind. For our present purposes we add the convention that $p_n^{(\infty)}(x) = x^n$, i.e., the expansion of a function in terms of $p_n^{(\infty)}$, is its Taylor expansion about the origin. Szegö [1, p. 95] shows that for $0 < \lambda < \infty$

$$p_n^{(\lambda)}(x) = \sum_{j=0}^{n}{}' a_{j,n}^{(\lambda)} T_j(x) \qquad (3.78)$$

with

$$a_{j,n}^{(\lambda)} \geqslant 0, \qquad j = 0, \ldots, n; \ n = 0, 1, 2, \ldots; \ a_{0,n}^{(\lambda)} + a_{1,n}^{(\lambda)} > 0, \qquad (3.79)$$

and (3.79) also holds when $\lambda = \infty$ according to Exercise 1.5.31. An obvious consequence of (3.78) and (3.79) is that for $\lambda > 0$

$$|p_n^{(\lambda)}(x)| \leqslant p_n^{(\lambda)}(1) = 1, \qquad -1 \leqslant x \leqslant 1. \qquad (3.80)$$

Let

$$s_{n,\lambda}(f; x) = \sum_{k=0}^{n} f_k(\lambda) \, p_k^{(\lambda)}(x), \qquad n = 0, 1, 2, \ldots,$$

be the partial sums of the expansion of f in terms of the $p_k^{(\lambda)}$, and let us put

$$\max_{-1 \leqslant x \leqslant 1} |f(x) - s_{n,\lambda}(f; x)| = \|f - s_{n,\lambda}(f)\| = S_n(f; \lambda).$$

THEOREM 3.17 (Rivlin and Wilson [1])

If $\lambda > 0, f_k(\lambda) \geqslant 0$ for $k > n$, and

$$\sum_{k=0}^{\infty} f_k(\lambda) \qquad (3.81)$$

converges, then

$$S_n(f; \lambda) \geqslant S_n(f; 0) = S_n(f). \qquad (3.82)$$

Equality holds in (3.82) only if $f \in \mathcal{P}_n$.

Proof

$$S_n(f; \lambda) = f(1) - s_{n,\lambda}(1) = \sum_{k=n+1}^{\infty} f_k(\lambda). \qquad (3.83)$$

Now,

$$A_j(f) = \frac{2}{\pi} \int_{-1}^{1} f(x) \, T_j(x) \, \frac{dx}{\sqrt{1 - x^2}}$$

$$= \frac{2}{\pi} \int_{-1}^{1} \left[\sum_{k=0}^{\infty} f_k(\lambda) \, p_k^{(\lambda)}(x) \right] T_j(x) \, \frac{dx}{\sqrt{1 - x^2}}$$

$$= \sum_{k=0}^{\infty} f_k(\lambda) \left[\frac{2}{\pi} \int_{-1}^{1} p_k^{(\lambda)}(x) \, T_j(x) \, \frac{dx}{\sqrt{1 - x^2}} \right]$$

$$= \sum_{k=j}^{\infty} f_k(\lambda) \, a_{j,k}^{(\lambda)}, \qquad (3.84)$$

where the $a_{j,k}^{(\lambda)}$ are defined in (3.78) and the term-by-term integration is justified by the convergence of (3.81), bearing in mind our normalization of the $p_k^{(\lambda)}$. Also, by (3.78) we have

$$\sum_{j=0}^{n}{}' a_{j,k}^{(\lambda)} = 1;$$

hence

$$\sum_{k=n+1}^{\infty} f_k(\lambda) = \sum_{k=n+1}^{\infty} f_k(\lambda) \cdot \sum_{j=0}^{n}{}' a_{j,k}^{(\lambda)} = \sum_{j=0}^{\infty}{}' \sum_{k=n+1}^{\infty} f_k(\lambda) a_{j,k}^{(\lambda)}$$

$$= \sum_{j=0}^{n}{}' \left[\sum_{k=n+1}^{\infty} f_k(\lambda) a_{j,k}^{(\lambda)} \right] + \sum_{j=n+1}^{\infty} \left[\sum_{k=j}^{\infty} f_k(\lambda) a_{j,k}^{(\lambda)} \right]$$

$$= C + \sum_{j=n+1}^{\infty} A_j(f), \qquad (3.85)$$

in view of (3.84), with $C \geqslant 0$. Since $A_j(f) \geqslant 0$ for $j = n + 1, \ldots$, by (3.84), we have

$$\sum_{j=n+1}^{\infty} A_j(f) = S_n(f; 0)$$

and the first part of the theorem is proved.

If $f \notin \mathscr{P}_n$ so that, say, $f_m(\lambda) > 0$ for $m > n$, then, since either $a_{0,m}^{(\lambda)}$ of $a_{1,m}^{(\lambda)}$ is positive according to (3.79), either $f_m(\lambda) a_{0,m}^{(\lambda)}$ or $f_m(\lambda) a_{1,m}^{(\lambda)}$ is positive and the quantity C in (3.85) is positive, which implies that the inequality holds in (3.82). ∎

Remark 1.   As a matter of fact (3.82) can be replaced by

$$S_n(f; \mu) \geqslant S_n(f; 0), \qquad 0 \leqslant \mu \leqslant \lambda. \qquad (3.86)$$

To see this we need the information (cf. Askey [1], but correct a misprint by reversing the inequality on the fourth line from the bottom of p. 301, and Rainville [1] for the case $\lambda = \infty$) that if $-(\tfrac{1}{2}) < \mu \leqslant \lambda$ and

$$p_n^{(\lambda)}(x) = \sum_{j=0}^{n} a_{j,n}^{(\lambda,\mu)} p_n^{(\mu)}(x)$$

then

$$a_{j,n}^{(\lambda,\mu)} \geqslant 0, \qquad j = 0, \ldots, n; \ n = 0, 1, 2, \ldots.$$

Thus mimicking (3.84) yields

$$f_j(\mu) = \sum_{k=j}^{\infty} f_k(\lambda) a_{j,k}^{(\lambda,\mu)}, \tag{3.87}$$

hence $f_k(\lambda) \geqslant 0$ for $k > n$ implies $f_k(\mu) \geqslant 0$ for $k > n$, and the theorem can be applied with λ replaced by μ, provided that

$$\sum_{k=0}^{\infty} f_k(\mu) < \infty.$$

This follows from the analog of (3.85) (with $a_{j,k}^{(\lambda)}$ replaced by $a_{j,k}^{(\lambda,\mu)}$ and so on).

Remark 2. The theorem remains true if the coefficients $f_k(\lambda)$ alternate in sign for $k > n$. To see this we observe that the $p_k^{(\lambda)}$ are even functions for even k and odd functions for odd k; thus applying the theorem to $f(-x)$ proves the analogous result for alternating coefficients.

Remark 3. If the coefficients are neither positive nor alternate in sign, (3.86) need not hold, as the example $f(x) = x^3 - (\frac{3}{2})x^2 - x; \lambda = \infty; n = 0$, shows. We have for $0 \leqslant \alpha < \infty$

$$f_0(\alpha) = \frac{\int\limits_{-1}^{1} f(x)(1 - x^2)^{\alpha - 1/2} \, dx}{\int\limits_{1}^{1} (1 - x^2)^{\alpha - 1/2} \, dx}$$

$$= \frac{-\frac{3}{2} \int\limits_{-1}^{1} x^2 (1 - x^2)^{\alpha - 1/2} \, dx}{\int\limits_{-1}^{1} (1 - x^2)^{\alpha - 1/2} \, dx} = -\frac{3}{4} \frac{1}{1 + \alpha}.$$

A small calculation gives

$$S_0(f; 0) = .89 \ldots,$$
$$S_0(f; \infty) = 1.5,$$
$$S_0(f; 0.1) = .82 \ldots.$$

4

ITERATIVE PROPERTIES

One of the most remarkable properties of the Chebyshev polynomials is the *semigroup* property (cf. Exercise 1.1.6)

$$T_m(T_n(x)) = T_{mn}(x).$$

An immediate consequence of the semigroup property is that the Chebyshev polynomials commute under composition; i.e.,

$$T_m(T_n) = T_n(T_m).$$

The first section of this chapter is devoted to this property. In the second section we shall study the ergodic and mixing properties of the Chebyshev polynomials considered as transformations of I onto itself, a study in which the semigroup property plays a role.

4.1. Permutable Polynomials

Two polynomials, p and q, are called *permutable* if $p(q(x)) = q(p(x))$ for all x. If we adopt the notation $p \cdot q$ to indicate the composition $p(q(x))$, then p and q are permutable if $p \cdot q = q \cdot p$. If p and q are permutable, we shall also say that p commutes with q and, of course, q commutes with p. Composition satisfies the associative law

$$p \cdot (q \cdot r) = (p \cdot q) \cdot r.$$

We shall write $p^{\{n\}}$ for the n-fold composition $p \cdot p \cdots p$. Since

$$T_m(T_n(x)) = T_n(T_m(x)) = T_{mn}(x),$$

we see that any two Chebyshev polynomials are permutable. Our first result is that no polynomials other than Chebyshev polynomials can commute with a given T_n if $n \geqslant 2$.

THEOREM 4.1 (Bertram [1])

If $n \geqslant 2$ and the polynomial p of degree $k \geqslant 1$ commutes with T_n, then $p = T_k$ if n is even and $p = \pm T_k$ if n is odd.

Proof. We know (cf. Exercise 2.4.6) that $\pm T_m(x)$ are the only polynomial solutions of

$$(1 - x^2)(y')^2 = m^2(1 - y^2) \tag{4.1}$$

for $m > 0$.

The theorem is proved by showing that, if p commutes with T_n, $y = p$ satisfies (4.1) with $m = k$.

The polynomial

$$q(x) = (1 - x^2)(p'(x))^2 - k^2(1 - p^2(x))$$

is in $\mathscr{P}_{2k-1}$, since the coefficient of x^{2k} is zero, but

$$n^2 q \cdot T_n = n^2(1 - T_n^2)(p' \cdot T_n)^2 - n^2 k^2(1 - (p \cdot T_n)^2)$$
$$= (1 - x^2)(T_n')^2(p' \cdot T_n)^2 - k^2(1 - p^2)(T_n' \cdot p)^2,$$

where we have used the permutability of p and T_n and the fact that T_n satisfies (4.1) with $m = n$. Now,

$$(p' \cdot T_n)T_n' = (p \cdot T_n)' = (T_n \cdot p)' = (T_n' \cdot p)p',$$

hence

$$n^2 q \cdot T_n = (1 - x^2)(p')^2(T_n' \cdot p)^2 - k^2(1 - p^2)(T_n' \cdot p)^2$$
$$= (T_n' \cdot p)^2((1 - x^2)(p')^2 - k^2(1 - p^2))$$
$$= (T_n' \cdot p)^2 q. \tag{4.2}$$

Suppose that $q \neq 0$ has degree s ($\leqslant 2k - 1$), then (4.2) implies that $sn = 2(n-1)k + s$ so that $s = 2k > 2k - 1$, a contradiction. Thus q is identically zero and $p = \pm T_k$. If n is even, $T_n \cdot (-T_k) = T_n \cdot T_k = T_k \cdot T_n \neq -T_k \cdot T_n$, hence $p = T_k$. If n is odd, $T_n \cdot (-T_k) = -T_n \cdot T_k = -T_k \cdot T_n$, hence $p = \pm T_k$. ∎

A sequence of polynomials, each of positive degree, containing at least one of each positive degree and such that every two polynomials in it are permutable is called a *chain*. The Chebyshev polynomials $T_1(x), \ldots, T_n(x), \ldots$, form a chain. So do the powers $\pi_j(x) \equiv x^j, j = 1, 2, \ldots$, as is easily verified. We shall see that these are essentially the only chains.

Suppose that

$$\lambda(x) = ax + b, \qquad a \neq 0, \tag{4.3}$$

so that

$$\lambda^{-1}(x) = \frac{x - b}{a} .$$

If p and q commute, it is clear that $\lambda^{-1} \cdot p \cdot \lambda$ and $\lambda^{-1} \cdot q \cdot \lambda$ also commute. Thus for any λ of the form (4.3) the sequences $\lambda^{-1} \cdot T_j \cdot \lambda, j = 1, 2, \ldots$, and $\lambda^{-1} \cdot \pi_j \cdot \lambda$, $j = 1, 2, \ldots$, are also chains, and this is the reason the word "essentially" was needed above. We shall say that p and $\lambda^{-1} \cdot p \cdot \lambda$ are *similar*; hence our goal is to show that the sequences $\{T_j\}$ and $\{\pi_j\}$ are the only chains, up to similarities. A first step in this direction is a companion piece to Theorem 4.1.

THEOREM 4.2

If $n \geqslant 2$ and the polynomial p of degree $k \geqslant 1$ commutes with $\pi_n(x)(=x^n)$ then $p = \pi_k$ if n is even and $p = \pm \pi_k$ if n is odd.

Proof. $y = \pi_n(x)$ satisfies

$$xy' = ny. \tag{4.4}$$

The polynomial $q(x) = x\, p'(x) - k\, p(x)$ is in $\mathscr{P}_{k-1}$, since the coefficient of x^k is zero. An argument analogous to that given in the proof of Theorem 4.1 yields $nq \cdot \pi_n = (\pi_n' \cdot p)q$, and if q is of degree $s \geqslant 0$ then $sn = k(n-1) + s$ implies that $s = k$, a contradiction; q must therefore be the zero polynomial. Hence $y = p$ satisifies (4.4) with n replaced by k, which means that $p(x) = cx^k (c \neq 0)$. The requirement that p commute with π_n implies that $cx^{kn} = c^n x^{kn}$, i.e., $c^{n-1} = 1$. Since c must be real, $c = 1$ if n is even and $c = \pm 1$ if n is odd.   ∎

THEOREM 4.3

There is at most one polynomial of degree $k \geqslant 1$ permutable with a given quadratic, $s(x) = a_0 + a_1 x + a_2 x^2, a_2 \neq 0$.

Proof.   If we put

$$\lambda(x) = \frac{x}{a_2} - \frac{a_1}{2a_2}, \tag{4.5}$$

we obtain

$$(\lambda^{-1} \cdot s \cdot \lambda)(x) = x^2 + c,$$

where $c = a_0 a_2 + (a_1/2) - (a_1{}^2/4)$. Thus to prove the theorem it suffices to show that there cannot be two distinct polynomials of degree k commuting with $x^2 + c$, for, if f and g are distinct polynomials of degree k commuting with s, there are distinct polynomials of degree k similar to f and g via (4.5) that commute with $x^2 + c$.

Suppose that p and q are distinct polynomials that satisfy

$$p(x^2 + c) = p^2(x) + c,$$
$$q(x^2 + c) = q^2(x) + c; \tag{4.6}$$

then comparing leading coefficients on both sides of each equality reveals that p and q both have leading coefficient 1. Thus $r = p - q \in \mathscr{P}_{k-1}$. Also

$$r(x^2 + c) = p^2(x) - q^2(x) = r(x)(p(x) + q(x)). \tag{4.7}$$

If the degree of r is $t \geqslant 0$, then according to (4.7) $2t = t + k$ or $t = k$, a contradiction. Therefore r is the zero polynomial and $p = q$. This contradiction establishes the theorem. ∎

An immediate consequence of Theorem 4.3 is that a chain contains exactly one polynomial of each positive degree; i.e., a chain is a sequence $\{p_j\}, j = 1, 2, \ldots,$ where p_j is of degree j and each pair of polynomials commutes. Two chains are called *similar* if there exists a $\lambda(x)$ satisfying (4.3) such that each polynomial in one is similar via λ to the polynomial of the other of the same degree. We can now prove our main result.

THEOREM 4.4

Every chain is either similar to $\{x^j\}, j = 1, 2, \ldots,$ *or* $\{T_j\}, j = 1, 2, \ldots.$

Proof. Let $\{p_j\}, j = 1, 2, \ldots,$ be a chain, with $p_2(x) = a_0 + a_1x + a_2x^2$. Let $\{q_j\}, j = 1, 2, \ldots,$ be the chain similar to $\{p_j\}$ with λ as defined in (4.5). Then $q_2(x) = x^2 + c; q_3$ commutes with q_2, hence

$$q_3(x^2 + c) = q_3^2(x) + c. \tag{4.8}$$

Thus $q_3^2(-x) = q_3^2(x)$, and since q_3 is of degree 3 we see that $q_3(-x) = -q_3(x)$; i.e., q_3 is an odd polynomial, say,

$$q_3(x) = b_1x + b_3x^3. \tag{4.9}$$

If we substitute (4.9) into (4.8) and equate coefficients of like powers, we obtain $b_3 = 1, b_1 = (\frac{3}{2})c,$

$$c(c + 2) = 0$$

and

$$c(2 + c)(2c - 1) = 0.$$

Therefore the only possible values of c are -2 and 0. If $c = 0$, then $q_2(x) = x^2$ and, according to Theorem 4.2, $q_j(x) = x^j$ for $j = 1, 2, \ldots,$ and $\{p_j\}$ is similar to $\{x^j\}$.

If $c = -2$ consider the chain $\{\mu^{-1} \cdot q_j \cdot \mu\}$, where $\mu(x) = 2x$. Since

$$(\mu^{-1} \cdot q_2 \cdot \mu) = T_2,$$

Theorem 4.1 informs us that

$$\mu^{-1} \cdot q_j \cdot \mu = T_j, \qquad j = 1, 2, \ldots .$$

Thus $\{p_j\}$ is similar to $\{T_j\}$ via the linear transformation $\lambda \cdot \mu$. ∎

This theorem is proved by Block and Thielman [1] and Jacobsthal [1], and the proof given here is an amalgam of their work.

EXERCISES 4.1

4.1.1.   If p commutes with q_1 and q_2, it commutes with $q_1 \cdot q_2$ and $q_2 \cdot q_1$.

4.1.2.   If q_1 and q_2, each of positive degree, commute with the same polynomial of degree 2, they are permutable.

A set of polynomials is called a P-set if every two polynomials are permutable. A P-set is *closed* if together with p and q it contains $p \cdot q$; it is called *complete* if no polynomial of positive degree that is not in the set commutes with all members of the set.

4.1.3.   A complete P-set is closed.

4.1.4.   A chain is a complete P-set.

4.1.5. p commutes with every polynomial if, and only if, $p = x$.

4.1.6. $\{x + a\}$ where a runs over the reals is a complete P-set.

4.1.7. $\{t(x - \alpha) + \alpha\}$ where α is fixed and t runs over the reals is a complete P-set.

4.1.8.   If p of degree 2 and q of degree 3 are permutable, then p and q are similar, via a common linear transformation, to either x^2 and x^3 or T_2 and T_3.

4.1.9.   If p of degree 2 and q of degree 4 are permutable, then $q = p \cdot p$.

A complete description can be given of permutable polynomials. Julia [1] and Ritt [1] showed that if p and q commute either both are iterates of the same polynomial or both are similar, with respect to the same λ, to either Chebyshev polynomials or powers. Thus Theorem 4.4 is an immediate consequence of this definitive result. Unfortunately the methods of Julia and Ritt are formidably complicated and we cannot present them here. What we *shall* do next is give a complete description of all polynomials that commute with a given quadratic, a task that is amenable to elementary analysis.

In view of the proof of Theorem 4.3, it suffices to consider polynomials p that commute with $x^2 - c$. Theorem 4.3 tells us that there is at most one polynomial of degree $k \geqslant 1$ that commutes with $x^2 - c$. If p, of degree k, commutes with $x^2 - c$, then

$$p(x^2 - c) = p^2(x) - c \tag{4.10}$$

and the leading coefficient of p is 1. Also, if we replace x by $-x$ in (4.10), we obtain $p(x^2 - c) = p^2(-x) - c$, hence $p^2(x) = p^2(-x)$ or $p(x) = \pm p(-x)$. Thus p is even if k is even and odd if k is odd. If $k = 1, p = x$, which, of course, commutes with every polynomial. Putting this trivial case aside, we wish to establish the Julia–Ritt result in the special case that (4.10) holds with $k \geqslant 2$. To this end some Lemmas are useful.

LEMMA 4.1.1

If $c < 0$ or $c > 2$, the sequence defined by

$$t_{n+1} = (t_n - c)^2, \qquad n = 1, 2, \ldots, \tag{4.11}$$

with $t_1 = c^2$, is strictly monotone increasing.

Proof. We note first that (4.11) implies that

$$t_2 = (c^2 - c)^2 = c^2(c - 1)^2 > c^2 = t_1.$$

Next we claim that $t_n > c^2$ for $n \geqslant 2$. As we have just seen, this is the case for $n = 2$. Suppose that $t_k > c^2$.

$$t_{k+1} - (t_k - c)^2 - t_k(t_k - 2c) + c^2.$$

Since $t_k > c^2, t_k - 2c > c^2 - 2c = c(c - 2) > 0$; hence $t_{k+1} > c^2$ and by mathematical induction $t_n > c^2$ for $n \geqslant 2$. Finally, $t_{n+1} > t_n, n = 1, 2, \ldots$. The case of $n = 1$ has been established. Suppose $t_k > t_{k-1}$.

$$t_{k+1} - t_k = (t_k - c)^2 - (t_{k-1} - c)^2 = (t_k - t_{k-1})(t_k + t_{k-1} - 2c),$$

but

$$(t_k + t_{k-1} - 2c) > 2c^2 - 2c = 2c(c - 1) > 0;$$

thus $t_{k+1} - t_k > 0$ and the lemma is established.   ∎

LEMMA 4.1.2

If $0 < c < 2$, the sequence defined by

$$t_{n+1} = \sqrt{t_n} + c, \qquad n = 1, 2, \ldots, \tag{4.12}$$

with $t_1 = c^2$, is strictly monotone increasing and $t_n > c^2$ for $n \geqslant 2$.

Proof.   We have

$$t_2 = 2c > c^2 = t_1.$$

We claim that $t_n \geq 2c, n \geq 2$. This is true for $n = 2$. Suppose it is true for $n = k$. Since $2c > c^2$, we see that $\sqrt{2c} > c$ and $\sqrt{2c} + c > 2c$. If $t_k \geq 2c$, then

$$t_{k+1} = \sqrt{t_k} + c \geq \sqrt{2c} + c > 2c,$$

establishing our claim by mathematical induction. Hence $t_n > c^2, n \geq 2$. Suppose next that $t_k > t_{k-1}$. We have

$$t_k = (t_{k+1} - c)^2,$$
$$t_{k-1} = (t_k - c)^2,$$

and therefore

$$0 < t_k - t_{k-1} = (t_{k+1} - c)^2 - (t_k - c)^2$$
$$= (t_{k+1} - t_k)(t_{k+1} + t_k - 2c).$$

We have seen that $t_k \geq 2c$; hence $t_{k+1} + t_k > 2c$ and $t_{k+1} - t_k > 0$. Since $t_2 > t_1$, the strict monotone increase of $\{t_n\}$ is established. ■

THEOREM 4.4

If $p(x)$, a polynomial of degree $k \geq 2$, commutes with $x^2 - c$, then either $p = x^k$ or $p = 2T_k(x/2)$ or p is an iterate of $(x^2 - c)$.

Proof.   (i) Suppose that $k = 2m - 1, m \geq 2$. Then p is odd and we can write

$$p(x) = xq(x^2),$$

where q is of degree $m - 1$. Equation 4.10 implies that

$$(x^2 - c)q((x^2 - c)^2) = x^2 q^2(x^2) - c,$$

and if we put $x^2 = t$

$$(t - c)\, q((t - c)^2) = tq^2(t) - c. \tag{4.13}$$

Suppose that $c < 0$ or $c > 2$ and $\{t_n\}$ is defined by (4.11). We claim that

$$q(t_n) = 1, \quad n = 1, 2, \dots . \tag{4.14}$$

Putting $t = 0$ in (4.13) yields $-cq(c^2) = -c$, and since $c < 0$

$$q(c^2) = 1, \tag{4.15}$$

establishing our claim for $n = 1$. Suppose that $q(t_k) = 1$. Putting $t = t_k$ in (4.13) yields

$$(t_k - c)\, q((t_k - c)^2) = t_k q^2(t_k) - c$$

or

$$t_{k+1}^{1/2}\, q(t_{k+1}) = t_k - c = t_{k+1}^{1/2}.$$

Since, in view of Lemma 4.1.1, $t_{k+1} > 0$, we obtain

$$q(t_{k+1}) = 1.$$

Thus (4.14) is proved by mathematical induction. Therefore $q(t)$ takes on the value 1 at least at m distinct (distinct, since monotone increasing by Lemma 4.1.1) points $t_1, \ldots, t_m$, hence is identically 1, contradicting the fact that it is of degree $m - 1 \geqslant 1$. Thus we must have $0 \leqslant c \leqslant 2$.

Suppose then that $0 < c < 2$. Now let $\{t_n\}$ be defined by (4.12). We claim that again $q(t_n) = 1$, $n = 1, \ldots$. In view of (4.15), this is true for $n = 1$. Suppose $q(t_k) = 1$. Then

$$(t_{k+1} - c)\, q((t_{k+1} - c)^2) = t_{k+1} q^2(t_{k+1}) - c$$

or

$$t_k^{1/2} = (t_k^{1/2} + c)\, q^2(t_{k+1}) - c.$$

Since $t_k^{1/2} + c = t_{k+1} > 0$, we obtain

$$q^2(t_{k+1}) = 1.$$

If $q(t_{k+1}) = -1$, putting $t_{k+2} = t$ in (4.13) yields

$$-(t_{k+2} - c) = t_{k+2} q^2(t_{k+2}) - c \geqslant -c$$

or $t_{k+1}^{1/2} \leqslant c$ and $t_{k+1} \leqslant c^2$, contradicting Lemma 4.1.2. Thus $q(t_{k+1}) = 1$, and by mathematical induction we have shown that $q(t_n) = 1$, $n = 1, \ldots, m$, where according to Lemma 4.1.2 $t_1 < \cdots < t_m$. Therefore $q = 1$, contradicting the fact that the degree of q is at least 1.

The only possible values of c are therefore seen to be 0, 2. If $c = 0$, then $p = x^k$ according to Theorem 4.2, whereas, if $c = 2$, $x^2 - 2 = 2T_2(x/2)$ and so $p = 2T_k(x/2)$ certainly commutes with $x^2 - 2$. This concludes our proof in the case of odd k.

(ii) Suppose that $k = 2m$, $m \geqslant 1$. If $2m = 2^s$, then $p = (x^2 - c)^{\{s\}}$ (the sth iterate of $x^2 - c$). Suppose that

$$2m = 2^s l,$$

where $l \geqslant 3$ is odd. Since p is an even function,

$$p_1(x) = p(\sqrt{x + c})$$

is a polynomial of degree $2^{s-1} l$ which satisfies

$$p_1(x^2 - c) = p_1^2(x) - c;$$

i.e., it commutes with $x^2 - c$. If $s = 1$, p_1 is of odd degree l, and therefore $c = 0, 2$ according to (i) above. If $s > 1$, p is of even degree, hence an even function, and $p_2(x) = p_1(\sqrt{x + c})$ commutes with $x^2 - c$ and is of degree $2^{s-1}l$. Continuing in this fashion, we see that $p_s(x)$ is of degree l and commutes with $x^2 - c$. Therefore $c = 0, 2$ and we conclude as in part (i).   ■

4.2. Ergodic and Mixing Properties

The Chebyshev polynomial $T_n(x)$ defines a mapping

$$x \to T_n(x)$$

of I (the interval $[-1, 1]$) onto I for each $n = 1, 2, \ldots$, which we denote by

$$T_n : I \to I. \tag{4.16}$$

Under this mapping each point of I, except ± 1, is the image of n distinct points of I, since the mapping

$$T_n : (\eta_i^{(n)}, \eta_{i-1}^{(n)}) \to (-1, 1), \qquad i = 1, \ldots, n$$

is one-to-one and onto. The mapping inverse to T_n is written T_n^{-1} and is an n-valued mapping except at ± 1. The effect of the mapping (4.16) onto the sub-interval $[0, \frac{1}{2}]$ of I for $n = 5$ is shown schematically in Figure 4.1.

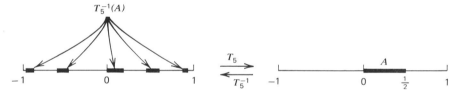

Fig. 4.1

The questions we shall be answering concern "metric" and "mixing" properties of the mapping T_n^{-1}, the sequence of mappings $T_1^{-1}, \ldots, T_n^{-1}, \ldots$, and the iterates T_n^{-k} (meaning here and henceforth the k-fold composition of T_n^{-1}). We shall try to make precise the vague notions that the image intervals of A in Figure 4.1 under T_5^{-1} combined have the same "length" as A and that the mappings $T_n^{-1}A$, $n = 1, 2, \ldots$, and $T_5^{-k}A$, $k = 1, 2, \ldots$, increasingly homogenize or mix A throughout I.

The vocabulary appropriate for making these notions precise is that of measure theory (cf. Halmos [1]). Let $(X, \mathscr{B}, \mu)$ be a separable finite measure space and let τ be a mapping of X onto itself that is measurable, i.e., such that $B \in \mathscr{B}$ implies $\tau^{-1}B \in \mathscr{B}$. τ is said to be *measure preserving* if

$$\mu(\tau^{-1}B) = \mu(B), \qquad B \in \mathscr{B}; \tag{4.17}$$

and if τ is measure preserving it is called *strongly mixing* if

$$\lim_{k \to \infty} \mu(\tau^{-k}A \cap B) = \frac{\mu(A)\,\mu(B)}{\mu(X)} \tag{4.18}$$

for all $A, B \in \mathcal{B}$. Every strongly mixing transformation is *ergodic*; i.e., if

$$\tau^{-1}A = A \tag{4.19}$$

for some $A \in \mathcal{B}$, then either $\mu(A) = 0$ or $\mu(A) = \mu(X)$, for if (4.19) holds (4.18) implies that

$$\mu(A \cap B) = \frac{\mu(A)\,\mu(B)}{\mu(X)}$$

for all $B \in \mathcal{B}$. Hence, if $B = A$,

$$\mu(A)\left(\frac{\mu(A)}{\mu(X)} - 1\right) = 0,$$

and either $\mu(A) = 0$ or $\mu(A) - \mu(X)$.

Finally, a sequence $\tau_1, \tau_2, \ldots, \tau_n, \ldots$, of measurable transformations of X onto itself, each of which preserves the measure μ, is called *strongly mixing* if

$$\lim_{n \to \omega} (\tau_n^{-1}A \cap B) - \frac{\mu(A)\,\mu(B)}{\mu(X)} \tag{4.20}$$

for any $A, B \in \mathcal{B}$. The condition (4.17), that τ be measure preserving, has an equivalent functional form.

LEMMA 4.2.1

τ *is measure preserving if, and only if,*

$$\int_X f(\tau x)\,d\mu = \int_X f(x)\,d\mu \tag{4.21}$$

for all $f \in L^1(X, \mathcal{B}, \mu)$.

Proof.   (i) If (4.21) holds for all integrable f, then it holds when f is the *characteristic function* of $B \in \mathcal{B}$ (the characteristic function of a set has the value 1 on the set and the value 0 off the set). But when f is such

$$\int_X f(\tau x)\,d\mu = \mu(\tau^{-1}B)$$

and

$$\int_X f(x)\,d\mu = \mu(B).$$

(ii) Suppose that (4.17) holds. Then (4.21) holds when f is the characteristic function of any $B \in \mathscr{B}$, as we have just seen in (i). Similarly, (4.21) holds when f is a *simple function*, i.e., a function that takes on only finitely many distinct values and is therefore a finite linear combination of characteristic functions. If f is a nonnegative integrable function, then

$$\int_X f(x)\,d\mu = \sup\left[\int_X g(x)\,d\mu\colon g \text{ simple}; \quad 0 \leqslant g(x) \leqslant f(x)\right]$$

$$= \sup\left[\int_X g(\tau x)\,d\mu\colon g \text{ simple}; \quad 0 \leqslant g(x) \leqslant f(x)\right].$$

Now $g(x) \leqslant f(x)$ for all $x \in X$ if, and only if, $g(\tau x) \leqslant f(\tau x)$ for all $x \in X$; hence

$$\int_X f(x)\,d\mu = \sup\left[\int_X g(\tau x)\,d\mu\colon g \text{ simple}; \quad 0 \leqslant g(\tau x) \leqslant f(\tau x)\right]$$

$$= \sup\left[\int_X h(x)\,d\mu; h \text{ simple}; \quad 0 \leqslant h(x) \leqslant f(\tau x)\right]$$

$$= \int_X f(\tau x)\,d\mu,$$

and (4.21) is established for nonnegative integrable functions. Equation 4.21 is now seen to hold for any integrable function by writing it as a difference of its positive and negative parts. ■

We shall also need a functional form for strong mixing.

LEMMA 4.2.2

The sequence $\tau_1, \ldots, \tau_n, \ldots$, is strongly mixing with respect to the measure μ if, and only if,

$$\lim_{n \to \infty} \int_X f(\tau_n x)\,g(x)\,d\mu = \frac{1}{\mu(X)}\int_X f(x)\,d\mu \int_X g(x)\,d\mu \qquad (4.22)$$

for every $f, g \in L^2(X, \mathscr{B}, \mu)$.

Proof.   (i) If we take f to be the characteristic function of A and g to be the characteristic function of B, then (4.22) implies (4.20).

(ii) If (4.20) holds, then (4.22) is valid when f and g are characteristic functions of any $A, B \in \mathscr{B}$, respectively. Hence (4.22) also holds when f and g are simple functions and we recall that the simple functions are dense in $L^2(X, \mathscr{B}, \mu)$.

At this point we consider a more general situation. Let $h_0, h_1, \ldots, h_k, \ldots$, be functions in $L^2(X, \mathscr{B}, \mu)$ such that, given any $h \in L^2(X, \mathscr{B}, \mu)$ and $\epsilon > 0$, there exists

$$w = \sum_{i=0}^{m} c_i h_i$$

such that

$$\int_X (h(x) - w(x))^2 \, d\mu < \epsilon.$$

We shall show next that if (4.22) holds when $g = h_i, f = h_j$, for every $i = 0, 1, \ldots$, $j = 0, 1, \ldots$, then (4.22) holds for all $f, g \in L^2(X, \mathscr{B}, \mu)$. Choosing the h_i to be the appropriate characteristic functions $[(X, \mathscr{B}, \mu)$ is separable] then proves the lemma.

Suppose then that (4.22) holds when $g = h_i, f = h_j$, for every $i = 0, 1, \ldots$, $j = 0, 1, \ldots$; then it clearly holds when f and g are finite linear combinations, say u and v, of the h_i. Now suppose that f and g are any functions in $L^2(X, \mathscr{B}, \mu)$ and, given $\epsilon > 0$, u and v are finite linear combinations of the h_i such that

$$\int_X (f(x) - u(x))^2 \, d\mu < \epsilon^2; \qquad \int_X (g(x) - v(x))^2 \, d\mu < \epsilon^2. \tag{4.23}$$

We have

$$C = \int_X f(\tau_n x) g(x) \, d\mu - \frac{1}{\mu(X)} \int_X f(x) \, d\mu \int_X g(x) \, d\mu$$

$$= \left\{ \int_X [f(\tau_n x) - u(\tau_n x)] [g(x) - v(x)] \, d\mu + \int_X v(x) (f(\tau_n x) - u(\tau_n x)) \, d\mu \right.$$

$$\left. + \int_X u(\tau_n x) (g(x) - v(x)) \, d\mu \right\}$$

$$+ \left\{ \int_X u(\tau_n x) v(x) \, d\mu - \frac{1}{\mu(X)} \int_X u(x) \, d\mu \int_X v(x) \, d\mu \right\}$$

$$+ \left\{ \frac{1}{\mu(X)} \int_X u(x) \, d\mu \int_X v(x) \, d\mu - \frac{1}{\mu(X)} \int_X f(x) \, d\mu \int_X g(x) \, d\mu \right\}$$

$$= \{D\} + \{E\} + \{F\}.$$

We observe that (4.21) implies that

$$\int_X [f(\tau_n x) - u(\tau_n x)]^2 \, d\mu = \int_X [f(x) - u(x)]^2 \, d\mu$$

and

$$\int_X [u(\tau_n x)]^2 \, d\mu = \int_X [u(x)]^2 \, d\mu.$$

Thus Schwarz's lemma, together with (4.23), implies that $|D| < c_1 \epsilon$ for some constant c_1. Moreover, if n is sufficiently large $|E| < \epsilon$, since we have seen that the lemma is valid for u and v. Finally,

$$F = \frac{1}{\mu(X)} \left[\int_X (f - u) \, d\mu \int_X (g - v) \, d\mu - \int_X g \, d\mu \int_X (f - u) \, d\mu \right.$$
$$\left. - \int_X f \, d\mu \int_X (g - v) \, d\mu \right]$$

and Schwarz's lemma and (4.23) yield $|F| < c_2 \epsilon$ for some constant c_2. Thus $|C| < c_3 \epsilon$ for some constant c_3 and n sufficiently large, establishing the lemma. ∎

Let us turn now to the Chebyshev transformations.

THEOREM 4.5

Let $\mathscr{B}$ denote the family of Borel subsets of I, and let λ be Lebesgue measure. If μ is the measure defined by

$$\mu(B) = \frac{2}{\pi} \int_B \frac{\lambda(dx)}{\sqrt{1 - x^2}}, \qquad B \in \mathscr{B}, \tag{4.24}$$

then each $T_n, n = 1, 2, \ldots,$ preserves the measure μ.

Proof. Consider the measure space $(X', \mathscr{B}', \lambda')$, where X' is the interval $[0, \pi]$, $\mathscr{B}'$, the Borel subsets of X', and λ' is Lebesgue measure on $\mathscr{B}'$. Let R be the one-to-one measurable mapping of X onto X' defined by

$$R : x \to x' = \arccos x.$$

Put

$$V_n = R T_n R^{-1}.$$

If

$$\frac{k\pi}{n} \leqslant x' \leqslant \frac{(k+1)\pi}{n}, \qquad k = 0, 1, \ldots, n - 1,$$

we see that

$$V_n(x') = \begin{cases} nx' - k\pi, & k \text{ even} \\ -nx' + (k+1)\pi, & k \text{ odd} \end{cases}$$

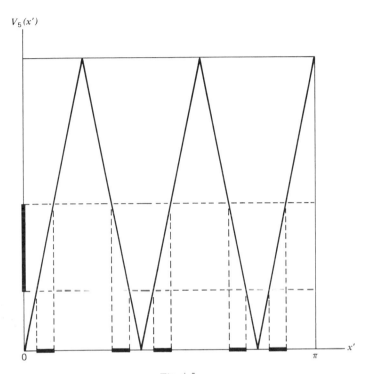

Fig. 4.2

$[V_5(x')$ is depicted in Figure 4.2]. An open subinterval of $[0, \pi]$ having length l is seen to be the image under V_n of n intervals, each of length l/n (as Figure 4.2 illustrates in the case $n = 5$). Thus V_n preserves Lebesgue measure. But if $-1 \leqslant a < b < 1$ then

$$\int_a^b \frac{dx}{\sqrt{1 - x^2}} = \int_{R(a)}^{R(b)} dx';$$

hence for $A \in \mathscr{B}$, $\mu(A) = (2/\pi) \lambda'(RA)$. Therefore $\mu(T_n^{-1}A) = (2/\pi) \lambda'(RT_n^{-1}A) = (2/\pi) \lambda'(RT_n^{-1}R^{-1}RA) = (2/\pi) \lambda'(V_n^{-1}RA) = (2/\pi) \lambda'(RA) = \mu(A).$ ∎

THEOREM 4.6 (Adler and Rivlin [1])

The sequence $T_1, T_2, \ldots, T_n, \ldots$, is strongly mixing with respect to the measure μ defined in (4.24).

Proof.   Suppose that $h \in L^2(X, \mathscr{B}, \mu)$. Let $s_k(h; x)$ be the partial sum of order k of the Chebyshev expansion of h (cf. p. 132). Familiar facts about Fourier series (cf. Zygmund [1, I, Chapter IV]) imply that, given $\epsilon > 0$, there exists k such that

$$\int_I [h(x) - s_k(h; x)]^2 \, d\mu < \epsilon.$$

If we recall the argument in part (ii) of the proof of Lemma 4.2.2 and choose $h_j = T_j(x)$, the theorem follows from Lemma 4.2.2 and the observation that, given any $i = 0, 1, \ldots, j = 0, 1, \ldots,$

$$\int_I T_i(T_n(x)) \, T_j(x) \, d\mu = \tfrac{1}{2} \int_I T_i(x) \, d\mu \int_I T_j(x) \, d\mu$$

for all sufficiently large n, in view of the semigroup and orthogonality properties of the Chebyshev polynomials. ■

COROLLARY 4.6.1

Each $T_n, n > 1$, is strongly mixing, hence ergodic.

Proof.   It is clear that the semigroup property implies that

$$T_n^{-k} = T_{nk}^{-1},$$

and so (4.18), with $\tau = T_n$, follows from the theorem.

As an amusing application of Theorem 4.6 we shall determine the limiting value as $n \to \infty$ of the area under the graph of $T_n(x)$ in the square with center at the origin and side 2 (see Figure 4.3). Let $K_n(y_2)$ be the area under the graph of $y = T_n(x)$ contained between the lines $x = -1, x = 1, y = -1, y = y_2$, where $-1 < y_2 \leqslant 1$. We shall establish the existence and determine the value of

$$K(y_2) = \lim_{n \to \infty} K_n(y_2).$$

Put $g(x) = (1 - x^2)^{1/2}$ and

$$f(x) = \begin{cases} x, & -1 \leqslant x \leqslant y_2, \\ y_2, & y_2 \leqslant x \leqslant 1. \end{cases}$$

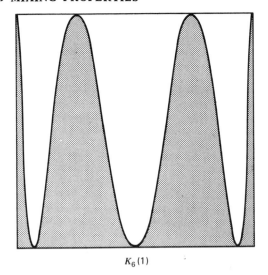

$$K_6(1)$$

Fig. 4.3

Then

$$K_n(y_2) = \int_{-1}^{1} (1 + f(T_n(x))) \, dx$$

$$= 2 + \frac{\pi}{2} \int_{I} f(T_n(x)) \, g(x) \, d\mu.$$

Thus, according to Theorem 4.6 and Lemma 4.2.2,

$$K(y_2) = \lim_{n \to \infty} K_n(y_2) = 2 + \frac{\pi}{2} \int_{I} f \, d\mu \int_{I} g \, d\mu.$$

Performing the integrations yields

$$K(y_2) = 2 + y_2 - \frac{2}{\pi} ((1 - y_2^2)^{1/2} + y_2 \arcsin y_2).$$

Thus $K(1) = 2$ and the limiting area under $T_n(x)$ as $n \to \infty$ is half the area of the square.

Moreover, by taking $g(x)$ to be the product of the characteristic function of $[x_1, x_2]$ and $(1 - x^2)^{1/2}$ we see that the limit as $n \to \infty$ of the area under $y = T_n(x)$

contained in the box $(-1\leqslant) x_1 \leqslant x \leqslant x_2(\leqslant 1), -1 \leqslant y \leqslant y_2$ is $(x_2 - x_1)K(y_2)/2$. Therefore the limit as $n \to \infty$ of the area under $y = T_n(x)$ bounded by the vertical lines $x = x_1$ and $x = x_2$, and the continuous curves $y = y_1(x)$ and $y = y_2(x)$, where $-1 \leqslant x_1 < x_2 \leqslant 1$ and $-1 \leqslant y_1(x) < y_2(x) \leqslant 1$ (for $x_1 \leqslant x \leqslant x_2$) is

$$\tfrac{1}{2} \int_{x_1}^{x_2} [K(y_2(x)) - K(y_1(x))] \, dx.$$

REFERENCES

Adler, R. L., and T. J. Rivlin

    1.   Ergodic and mixing properties of Chebyshev polynomials, *Proc. Amer. Math. Soc.*, **15** (1964), 794–796.

Askey, R.

    1.   Jacobi polynomial expansions with positive coefficients and imbeddings of projective spaces, *Bull. Amer. Math. Soc.*, **74** (1968), 301–304.

    2.   Mean convergence of orthogonal series and Lagrange interpolation, *Acta Math. Acad. Sci. Hungar.*, **23** (1972), 71–85.

    3.   A sine inequality, Problem 73–21, *SIAM Rev.*, **15** (1973), 788.

Askey, R., and G. Gasper, L. A. Harris

    1.   An inequality for Tchebycheff polynomials and extensions, *J. Approximation Theory*, to appear.

Berman, D. L.

    1.   Solution of an extremal problem in the theory of interpolation (Russian), *Dokl. Akad. Nauk SSSR*, **87** (1952), 167 170.

Bernstein, S. N.

    1.   Sur la meilleure approximation de $|x|$ par des polynomes de degrés donnés, *Acta Math.*, **37** (1913), 1–57.

    2.   *Leçons sur les Propriétés Extrémales et la Meilleure Approximation des Fonctions Analytiques d'une Variable Réelle*, Gauthier-Villars, Paris, 1926.

    3.   On a theorem of V. A. Markov (Russian), Trudy Leningrad Indust. Inst., No. 5, Phys.-math. Sci. Section, No. 1 (1938), 8–13. (Also in Vol. II of *The Collected Works of S. N. Bernstein, Akad. Nauk SSSR*, Moscow, 1954).

Bertram, E. A.

    1.   Polynomials which commute with a Tchebycheff polynomial, *Amer. Math. Monthly*, **78** (1971), 650–653.

Block, H. D., and H. P. Thielman

    1.   Commutative polynomials, *Quart. J. Math.* Oxford Ser. (2), **2** (1951), 241–243.

Boas, R. P., Jr.

    1.   Inequalities for the derivatives of polynomials, *Math. Mag.*, **42** (1969), 165–174.

Cavaretta, A. S., Jr.

    1.   An elementary proof of Kolmogorov's theorem, *Amer. Math. Monthly*, **81** (1974), 480–486.

De Vore, R. A.

1. A property of Chebyshev polynomials, *J. Approximation Theory*, to appear.

Duffin, R. J., and A. C. Schaeffer

1. A refinement of an inequality of the brothers Markoff, *Trans. Amer. Math. Soc.,* **50** (1941), 517–528.

Ehlich, H., and K. Zeller

1. Schwankung von Polynomen zwischen Gitterpunkten, *Math. Z.,* **86** (1964), 41–44.
2. Auswertung der Normen von Interpolationsoperatoren, *Math. Ann.,* **164** (1966), 105–112.
3. Čebyšev–Polynome in mehreren Veränderlichen, *Math. Z.,* **93** (1966), 142–143.

Erdös, P.

1. Problems and results on the theory of interpolation. II, *Acta Math. Acad. Sci. Hungar.,* **12** (1961), 235–244.

Erdös, P., and G. Grünwald

1. Note on an elementary problem of interpolation, *Bull. Amer. Math. Soc.,* **44** (1938), 515–518.

Faber, G.

1. Über die interpolatorische Darstellung stetiger Funktionen, *Jber. Deutsch. Math.- Verein.,* **23** (1914), 192–210.

Fejér, L.

1. Über Interpolation, *Gött. Nachr.* (1916), 66–91.
2. Die Abschätzung eines Polynoms in einem Intervalle, wenn Schranken fur seine Werte und ersten Ableitungswerte in einzelnen Punkten des Intervalles gegeben sind, und ihre Anwendung auf die Konvergenzfrage Hermitescher Interpolations-reihen, *Math. Z.,* **32** (1930), 426–457.
3. Bestimmung derjenigen Abszissen eines Intervalles, für welche die Quadratsumme der Grundfunktionen der Lagrangeschen Interpolation im Intervalle ein möglichst kleines Maximum besitzt, *Ann. Scuola, Norm. Sup. Pisa,* **1** (1932), 263–276.
4. Mechanische Quadraturen mit positiven Cotesschen Zahlen, *Math. Z.,* **37** (1933), 287–309.

Feldheim, E.

1. Sur l'orthogonalité des fonctions fondamentales et sur la forte convergence en moyenne des polynomes d'interpolation de Lagrange dans le cas des abscisses de Tchebychef, *Bull. Soc. Math. France,* **65** (1937), 1–40.

Fox, L., and I. B. Parker

1. *Chebyshev Polynomials in Numerical Analysis*, Oxford University Press, London, 1968.

Friedman, S. L.

1. Chebyshev constant and Chebyshev points, *Trans. Amer. Math. Soc.,* **186** (1973), 129–139.

Halmos, P. R.

1. *Measure Theory*, Van Nostrand, New York, 1950.

Hille, E.

1. Topics in classical analysis, *Lectures on Modern Mathematics*, Vol. III, T. L. Saaty, Ed. Wiley, New York, 1965.

Jacobsthal, E.

1. Über vertauschbare Polynome, *Math. Z.*, **63** (1955), 243–276.

Johnson, L. W., and R. D. Riess

1. Convergence of polynomials interpolating at the zeros of $T_n(x)$, *Math. Z.*, **116** (1970), 355–358.

Julia, G.

1. Mémoire sur la permutabilité des fractions rationnelles, *Ann. Sci. École Norm. Sup.*, **39** (1922), 131–215.

Karlin, S., and W. J. Studden

1. *Tchebycheff Systems: With Applications in Analysis and Statistics*, Wiley, New York, 1966.

Knopp, K.

1. *Theory and Application of Infinite Series*, Hafner, New York, 1949.

Korsak, A., and C. Schubert

1. A determinant expression of Tchebyshev polynomials, *Canad. math. Bull.*, **3** (1960), 243–246.

Lanczos, C.

1. *Applied Analysis*, Prentice-Hall, Englewood Cliffs, N. J., 1956.

La Vallée Poussin, C. J., de

1. *Leçons sur l'Approximation des Fonctions d'une Variable Réele*, Gauthier–Villars, Paris, 1952.

Lepson, B.

1. Upper bounds for Dirichlet kernels and Tchebycheff polynomials of the second kind, *J. Math. Anal. Appl.*, to appear.

Markov, V. A. (W. Markoff)

1. Über Polynome, die in einem gegeben Intervalle möglichst wenig von Null abweichen, *Math. Ann.*, 77 (1916), 213–258.

Matorin, A. P.

1. On inequalities between the maxima of the absolute values of a function and its derivatives on a half-line, *Amer. Math. Soc. Transl.*, Series 2, 8 (1958), 13–17.

Mohr, E.

1. Elementarer Beweis einer Ungleichung von W. A. Markoff, *Tensor (N. S.)*, **14** (1963), 71–85.

Natanson, I. P.

    1.   *Constructive Function Theory*, Vols. I, II, III, Ungar, New York: I, 1964; II, 1965; III, 1965.

Rainville, E. D.

    1.   *Special Functions*, Macmillan, New York, 1960.

Reimer, M.

    1.   Bounds for the Horner sums, *SIAM J. Numer. Anal.*, 5 (1968), 461–469.

Riesz, M.

    1.   Eine trigonometrische Interpolationsformel und einige Ungleichungen für Polynome, *Jber. Deutsch. Math.-Verein.*, 23 (1914), 354–368.

Ritt, J. F.

    1.   Permutable rational functions, *Trans. Amer. Math. Soc.*, 25 (1923), 399–448.

Rivlin, T. J.

    1.   *An Introduction to the Approximation of Functions*, Blaisdell, Waltham, Mass., 1969.

Rivlin, T. J., and H. S. Shapiro

    1.   A unified approach to certain problems of approximation and minimization, *J. Soc. Indust. Appl. Math.*, 9 (1961), 670–699.

Rivlin, T. J., and M. W. Wilson

    1.   An optimal property of Chebyshev expansions, *J. Approximation Theory, 2* (1969), 312–317.

Rogosinski, W. W.

    1.   Some elementary inequalities for polynomials, *Math. Gaz.*, 39 (1955), 7–12.

Rudin, W.

    1.   Some theorems on Fourier coefficients, *Proc. Amer. Math. Soc.*, 10 (1959), 855–859.

Shapiro, H. S.

    1.   *Extremal Problems for Polynomials and Power Series*, Master's thesis, MIT, 1951.

    2.   *Topics in Approximation Theory*, Lecture Notes in Mathematics, 187, Springer, New York, 1971.

Szegö, G.

    1.   *Orthogonal Polynomials*, American Mathematical Society, New York, 1959.

Tchakaloff, V.

    1.   Formules de cubatures mécaniques a coefficients non négatifs, *Bull. Sci. Math.* 2, 81 (1957), 123–134.

Tchebychef, P. L.

1. *Œuvres*, 2 vols., Chelsea, New York, 1962.

Titchmarsh, E. C.

1. *The Theory of Functions*, 2nd ed., Oxford University Press, London, 1939.

Turán, P.

1. On an inequality of Čebyšev, *Ann. Univ. Sci. Budapest. Eötvös Sect. Math.*, 11 (1968), 15–16.

Voronovskaja, E. V.

1. *The Functional Method and its Applications*, Translations of Mathematical Monographs, Vol. 28, American Mathematical Society, Providence, R. I., 1970.

Zygmund, A.

1. *Trigonometric Series*, Vols. I and II, Cambridge University Press, London, 1959.

GLOSSARY OF SYMBOLS

Symbol	Meaning	Page of first occurrence
B_n	Unit ball in $\mathscr{P}_n$	91
C_n	Convex subset of $\mathscr{P}_n$	92
$E(g;B)$	Subset of B on which $\lvert g \rvert = \lVert g \rVert$	61
$E_k(f)$	Error of best approximation by polynomials of degree at most k	11
$\eta_j,\ \eta_j^{(n)}$	Extrema of the Chebyshev polynomial of degree n	6
$h_j(x)$	Fundamental polynomials of the first kind for Hermite interpolation	24
$\mathfrak{h}_j(x)$	Fundamental polynomials of the second kind for Hermite interpolation	24
$l_{j,n}(x)$	Fundamental polynomials for interpolation	10
$L_k(f,X;x)$	Interpolating polynomial of degree k to f at the points of the $(k-1)$st row of X, evaluated at x.	11
$\lambda_k(X;x)$	Lebesgue function of polynomial interpolation theory	12
$\Lambda_k(X)$	Lebesgue constant of polynomial interpolation theory	12
$\omega(x)$	Polynomial with zeros at nodes	13
$\mathscr{P}_n$	The set of polynomials of degree at most n	1
$\phi_j,\ \phi_j^{(n)}$	Extrema of $\cos n\theta$	6
$\mathbb{R}^n$	Real n-space	57
$\hat{S}$	The convex hull of S	58
$s_j^{(n)}$	Partial sum of the coefficients of $T_n(x)$	9
$\mathscr{S}_k(z)$	The Shapiro polynomial of degree k.	129
σ	The domain or base of a signature	63
Σ	Signature	63
Σ'	Sum with the first term halved	35
Σ''	Trapezoidal sum	34
T	Array of nodes consisting of the zeros of the Chebyshev polynomial	13
$t_j^{(n)},\ t_j$	Coefficient of the Chebyshev polynomial of degree n	3

183

INDEX